CHEMICALS
in
ACTION

SECOND EDITION

Thomas R. Donovan

Marion C. Poole

Douglas J. Yack

Holt, Rinehart and Winston of Canada
a division of
Harcourt Brace & Company Canada, Ltd.

Toronto • Orlando • San Diego • London • Sydney

Canadian Cataloguing in Publication Data

Donovan, Thomas R
Chemicals in action

2nd ed.
Includes index.
ISBN 0-03-922455-4

1. Chemistry. 2. Chemistry—Problems, exercises, etc.
I. Poole, Marion C. II. Yack, Douglas J.
III. Title

QD33.D65 1994   540   C94-930015-2

Editorial Director: Hans Mills
Editor: Mike Wevrick
Senior Production Editor: Karin Fediw
Production Editors: Joe Piccione, Bob Douglas
Art Director: Arnold Winterhoff
Page Composition: Brian Lehen • Graphic Design
Illustrator: Terry May
Design: Clearlight Inc.
Photo Research: Liz Kirk
Cover Photograph by Ed Dosien. Reproduced with permission from
The Great Balloon Festival, © 1989, Free Flight Press, Inc.

∞ This book was printed in Canada on acid-free paper.

2 3 4 5   98 97

Contents

Preface

The second edition of *Chemicals in Action,* like the first edition, is an introductory course of study designed for the secondary level. It has two basic aims: to introduce students to some of the fundamental principles of chemistry; and to show students how chemistry is applied in their lives and in society.

Six years of popular use in the schools provided the authors and publishers with ideas from students and teachers on how *Chemicals in Action* could be improved. This expanded full-color edition has incorporated many of their suggestions.

Due, in part, to the rapid advances in applied science and technology, and, in part, to the rapid changes in society, many sections of the original edition have been revised and updated.

The truths of any science can be demonstrated only through extensive investigation and experimentation. To this end, *Chemicals in Action, second edition* still includes more than one hundred possible experiments and activities throughout the text. It presents an activity-oriented, hands-on approach.

Chemicals in Action, second edition is unique because it combines theory, applications, and the impact of chemical concepts on society. Theoretical knowledge is presented in a clear concise manner immediately preceding each experiment or activity related to the subject under discussion.

Activities bring the course to life, developing concepts and demonstrating their relevance. Many new activities have been added to the original version, some have been revised, and some of the original activities have been replaced by ones as equally interesting and relevant.

Each chapter begins with a short introductory passage designed to acquaint the students with the subject matter contained in the chapter and to arouse interest.

The introduction is followed by a series of **Key Ideas**. These Key Ideas include not only the objectives of the chapter but also other statements designed to heighten the students' awareness of the connection between science, technology, and the society in which they live.

The main concepts of each chapter are developed in short concise sections. The objectives of these sections are positively reinforced by a series of questions called **Try These,** which end most sections. The purpose of these questions is to reinforce theoretical concepts and to recall the main points of the material just read.

The main text of most chapters is followed by a **Career Opportunity** section, which features careers that can be pursued in a field related to the subject discussed in the chapter.

Just as the opening section of the chapter began with a list of Key Ideas, the concluding section of the chapter is accompanied by an alphabetical list of **Key Words**. This list can be used effectively by students to review the important terms and concepts contained in the chapter. Many of the Key Words are defined in the **Glossary** at the end of the text.

Science & Technology, Science & Society and **Science, Technology, & Society** sections have been added throughout this edition of the text. These sections present some of the interesting applications and extensions of the chemical concepts being developed in a chapter. Specific examples emphasize to the students the importance of becoming informed citizens.

Each chapter ends with a set of **Exercises**. The Review Exercises are designed to review the main ideas contained in the chapter and to improve student understanding of the concepts. Extension Exercises are designed to allow interested students the opportunity to investigate a topic in more depth. These can be used to develop higher level learning skills, such as problem-solving skills; to heighten student awareness of true-to-life dilemmas; and to increase language and research skills. Many of these questions are open-ended and, as such, have no right or wrong answers. They are designed to increase the student's ability to formulate informed opinions by viewing a particular situation from several perspectives.

Chemicals in Action, second edition, is written in a straightforward, easily understood style. The format is designed to create the impression of spaciousness in an open single-column format. Interesting photographs in full color serve to arouse interest and motivate as illustrations serve to clarify and add dimension. The text also makes use of colored type to highlight new words or ideas when these are introduced or defined.

In addition to the above, the factor that most affects a text's readability is its ability to motivate students to read it. As with the original edition, this is achieved in *Chemicals in Action, second edition* through the use of practical activities and problem-solving scenarios that will appeal to the student's sense of adventure, curiosity, and the desire for meaningful, relevant knowledge.

Acknowledgments

The authors would like to thank their spouses and families for their support and understanding during the many hours spent preparing this textbook.

The authors are also grateful to the many people who contributed their assistance. Several of our colleagues in the Science Teachers' Association of Ontario have inspired and helped us. In particular, we would like to thank those teachers who reviewed draft copies of the manuscript:

Ernestine Abraham, Youngstown, New York
Paul Barron, Ottawa, Ontario
Harvey D. Goodman, Queens, New York
Allison K. Leding, Cambridge, Minnesota
Jane Bray Nelson, Orlando, Florida
Richard A. Patz, Pompano Beach, Florida
Jim Read, Oshawa, Ontario
Peter H. Williams, Toronto, Ontario
Richard Woram, Timmins, Ontario
Douglas Wrigglesworth, York Region, Ontario
Thomas J. Yots, Youngstown, New York

Thanks are also due to the students of Mother Teresa High School who appear in many of the activity photographs.

To Mike Wevrick, Hans Mills, and the staff at Harcourt, Brace & Company, Canada, our special thanks. They have displayed unfailing dedication to the task and their help in the preparation of this program has been immense. We thank them for their patience, comments, suggestions, and improvements.

To the Student

Welcome to the second edition of *Chemicals in Action*. In this applied chemistry course, you will learn by doing. You will learn some of the fundamental principles of chemistry and how these principles affect your everyday life.

The key to *Chemicals in Action* is the laboratory. The laboratory activities give you the evidence for the concepts that are developed. These concepts will show you how the chemical part of the world works. Each unit in this program is organized in the same way. Each combines evidence, theory, applications of concepts, and their impact on society, including job possibilities.

Today, more than ever, science and technology are essential to the search for answers to problems. People want answers to questions about energy, pollution, and many other social issues. Often answers lead to more questions. For example, you will discover in this course how replacing soaps with detergents solved one problem but created another.

It is important for you to consider issues involving science, technology, and society. The answers to these questions will determine what the world you will live in will be like. Making decisions about many of these issues will require you to understand how science and technology work. That is the reason for this course.

Safety

As with all activities, care must be taken to avoid accidents in the school laboratory. Using proper equipment, materials, and procedures will help to keep the laboratory safe for everyone. Some general laboratory safety rules and symbols are listed in this section. Your teacher will provide you with further instructions that apply to your particular laboratory.

Safety Rules

1. Read each activity thoroughly before you begin. Always follow any safety precautions that are listed with the activity. Ask your teacher if you do not understand something in the instructions.

2. Dress appropriately for laboratory work. Tie back long hair. Do not wear loose or bulky clothing. Cotton is preferred to nylon or polyester. Do not wear long necklaces or heavy bulky jewelry. Shoes must be worn at all times.

3. Wear an apron and safety goggles (and gloves if specified) during each lab activity that involves potentially harmful chemicals. Do not wear contact lenses because of the possibility of chemicals getting underneath them.

4. Always stand during activities. Push chairs and stools under desks. Be careful not to bump into other students.

5. Learn fire safety precautions including fire exit routes, the location of the nearest fire alarm, and the location and use of the fire safety blanket.

6. Use tongs to handle hot equipment.

7. Never taste chemicals or touch them with your bare hands. Use a clean dry scoop or spatula to transfer chemicals. Clean and dry scoops and spatulas before using them for another chemical.

8. Be careful when lighting burners. If a burner does not light immediately, turn off the gas and wait a few minutes before trying again.

9. Keep your head and clothing away from the flame. Turn off the burner when you are done with it. Do not leave a lit burner unattended.

10. If you burn yourself, immediately run cold water over the burned area and have a classmate notify the teacher.

11. Use a clean glass lid when pouring a liquid from one container to another to prevent splashing.

12. Never add water to a chemical. Add the chemical to water.

13. If you spill chemicals on your hands or clothing, or in your eyes, rinse the chemicals off with plenty of water (unless instructions say not to).

14. Clean up spilled chemicals immediately.

15. Report all accidents, spills, or broken equipment to your teacher immediately, no matter how minor they seem.

16. Put any broken glass into the special "broken glass container," not the regular garbage.

17. Do not work alone in the laboratory or if your teacher is not present.

Safety Symbols

Caution notes and safety symbols have been used throughout the book to indicate specific safety precautions for each activity. The safety symbols and their meanings are as follows:

 Poison—keep away from your mouth

 Flammable—keep away from heat, sparks, and flames

 Explosive—do not heat or puncture

 Corrosive—do not spill on skin or clothes

 Radioactive—wear gloves, do not hold near your body

 Wear goggles and aprons

 Burner safety—see *Safety Rules* 8–10

 Waste disposal—dispose of waste chemicals as instructed by your teacher

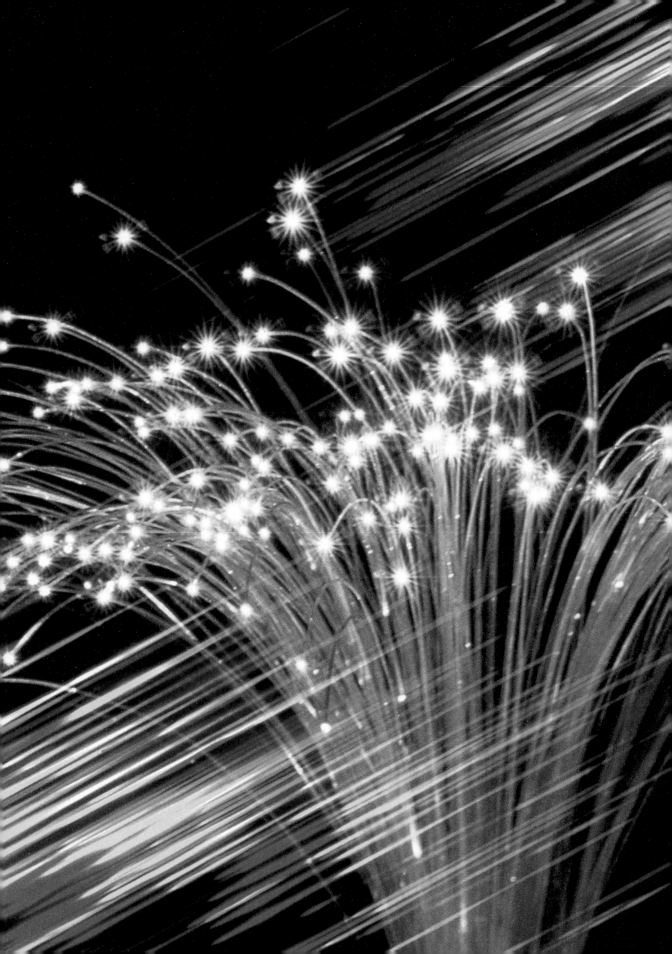

CHAPTER 1
WHAT'S IT ALL ABOUT?

In the last hundred years there has been an explosion of new materials and products that have dramatically changed people's lives. New materials and products are being developed as you read this.

These new items affect the types of clothes you wear, the hobbies you enjoy, and the food you eat. Synthetic fibers, video tapes, and diet soft drinks are just a few of the products that are available to you that were not available 50 years ago.

In this chapter, you will examine some aspects of applied chemistry and how they affect your life. Applied chemistry is that branch of science that uses basic chemical knowledge to develop products to improve people's lives.

Some of the people who use scientific knowledge to develop technologies are engineers and industrial professionals, often called technologists. They develop and test materials for uses that affect both everyday life and the future.

Key Ideas

- Applied chemistry is the use of chemical knowledge to solve problems and improve standards of living.

- Technology uses the basic knowledge of science to develop useful products for society.

- Chemists use the scientific method to explain phenomena.

- The technological method is used to make products to benefit society.

Fiber optics cables can transmit information much more rapidly and efficiently than copper wires.

1.1
TECHNOLOGY

Sir Alexander Fleming
(1881–1955), discoverer
of penicillin

It is almost impossible to exist in today's world without being affected by some application of chemistry. It is a science of invention and innovation.

Applied chemists take the laws and theories developed by pure chemists and apply them to some problem or need in society. The desired end result is a solution to the problem. This solution often is a new product, which is an example of the results of applied science or technology. Technology is another word for any kind of applied science, such as applied chemistry.

The 20th century has been marked by technological advances that have greatly benefited humanity in the fields of medicine, construction, communications, transportation, and energy production, just to name a few.

For example, in the field of medical technology, human illness and premature or early death have been dramatically decreased by the development of drugs such as penicillin. Many children died from dreaded diseases such as pneumonia before penicillin was developed by Sir Alexander Fleming of Scotland in 1929.

Advances in building-material technology have been dramatic. As recently as the 1800s, buildings were square or rectangular in shape and only three or four stories high. Advances in the types of metals, concrete, and plastics available have made it possible for architects to develop many new exciting designs. Today skyscrapers are common and are made of a wide variety of materials and in many different shapes.

Modern skyscrapers are one result of improvements in technology.

Communications technology also has experienced rapid advances. The 20th century has seen the development of television, computers, and communications satellites.

Fifty years ago, people were amazed to be able to see world events a day or two after they had happened. Today, via satellites in space, you can see events as they occur around the globe.

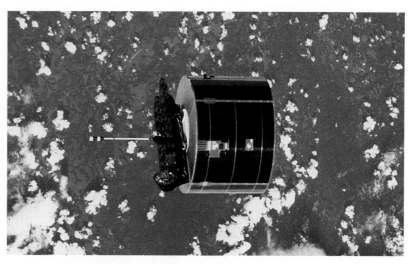

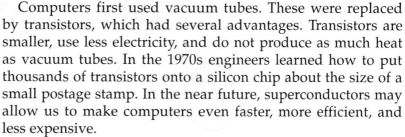

Communications satellites like this one allow information to be sent around the world very rapidly and without wires.

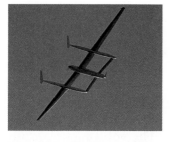

Use of low-density materials allowed *Voyager* to travel around the world without landing or refueling.

Computers first used vacuum tubes. These were replaced by transistors, which had several advantages. Transistors are smaller, use less electricity, and do not produce as much heat as vacuum tubes. In the 1970s engineers learned how to put thousands of transistors onto a silicon chip about the size of a small postage stamp. In the near future, superconductors may allow us to make computers even faster, more efficient, and less expensive.

Advances in transportation technology have been spectacular. In the near future, we may be able to use superconductors to travel on nearly frictionless trains.

In 1986 the airplane *Voyager* flew non-stop around the world in nine days. What was remarkable was that it never stopped to refuel. What made this feat possible was the use of low-density materials that reduced the amount of fuel required to complete the journey.

At the beginning of the 1900s, space flight was only a dream in a few people's minds. At that time, the construction materials, engines, fuels, and technology were not available. Today, space flights are almost routine. However, it was only in 1981 that the first reusable spacecraft, the *Columbia*, touched down on land.

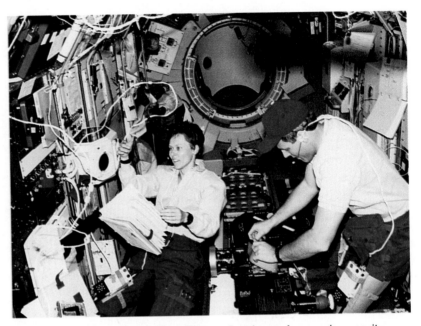

Canadian payload specialist Dr. Roberta Bondar performs microgravity experiments while U.S. astronaut Stephen S. Oswald changes the film on the IMAX camera aboard the space shuttle *Discovery*.

We depend on technology developed to produce energy from fossil fuels such as coal, oil, and natural gas for most of our energy, but supplies of these fuels are finite and burning them causes pollution.

However, we are already benefiting from one of the greatest recent technological achievements: the release of usable energy from the atom, nuclear power. The splitting of certain atoms of uranium in a nuclear power reactor is a source of heat used to produce steam to generate electricity. Both the United States and Canada have played important parts in the development of nuclear technology. You will learn more about nuclear power in Chapter 24.

Like fossil fuels, natural supplies of uranium will eventually be used up. However, nuclear fuels can be recycled to provide an additional source of energy. As well, researchers are trying to develop fusion reactors that will run on deuterium, a form of hydrogen that can be extracted from seawater. If this technology can be perfected, the world's energy concerns will be greatly reduced.

1.2
SCIENCE AND TECHNOLOGY

Many people think of science and technology as being the same thing. As you may know, they are not. There are two kinds of science: pure science and applied science. **Pure science** is the search for scientific knowledge. **Applied science** is the use of that knowledge to solve problems.

However, there are not always clear lines of distinction between the two sciences. In many cases, the pure scientist works on a well-defined problem. When this is the case, the pure scientist tries to provide the technologist with the specific information required to produce the desired end result. In other cases, a scientist may be involved in both the pure and the applied sciences related to a project.

Table 1.1   The two branches of science

Type of Science	Definition	Examples
pure	seeks knowledge	structure of matter, biochemistry of proteins
applied	uses knowledge	nuclear power, medicines

We live in a technological age. Prior to 1800, technology had produced relatively few products. For example, from prehistoric times until recently, cooking was done mostly over coal or wood fires. When natural gas was discovered, technologists saw possibilities to fulfill a need. They worked to find ways to pipe gas into homes. They also designed efficient and safe gas-burning stoves.

When scientists discovered that passing electricity through metal caused it to heat up, engineers designed and produced electric stoves and other heating appliances. One recent development in cooking technology is the microwave oven. Each invention saves us time and work. The **technological method** is used to develop products for our benefit.

A nuclear power station

The Technological Method

1. A problem or need requires a solution.
2. Scientists begin to experiment or do research to find the solution. (This step is not always included.)
3. An idea for a solution to the problem is imagined.
4. A plan or blueprint is made of the product to be developed.
5. The first type of its kind (a **prototype**) is produced.
6. The prototype is tested to see how well it performs as a solution to the problem or need.
7. Based on its performance, the prototype is changed and improved to remove any flaws or defects.
8. The finished product is marketed.

Try These

1. What is the difference between science and technology?
2. Identify a problem that you think will be solved by the technological method within the next 50 years. What do you imagine the prototype will be?
3. Name one product that you think is or was based on a good theory but a bad design, and one product that is an example of a good theory and a good design.

1.3
THE SCIENTIFIC METHOD

A chemist's approach to obtaining knowledge about matter is similar to that of any other scientist's. Some important discoveries come about quite by accident, some from ingenious ideas, and others from much hard work. Often all three factors are involved. Most scientific knowledge is the result of carefully planned experiments by highly trained scientists. The scientist's goal is to explain and predict natural phenomena.

The following chart summarizes the steps a scientist goes through when solving a problem. This sequence is called the **scientific method**

The Scientific Method

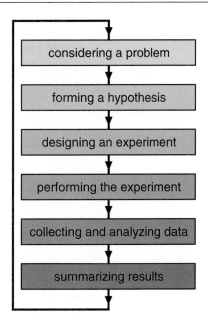

```
┌─────────────────────────────────┐
│   considering a problem         │
└─────────────────────────────────┘
              ↓
┌─────────────────────────────────┐
│   forming a hypothesis          │
└─────────────────────────────────┘
              ↓
┌─────────────────────────────────┐
│   designing an experiment       │
└─────────────────────────────────┘
              ↓
┌─────────────────────────────────┐
│   performing the experiment     │
└─────────────────────────────────┘
              ↓
┌─────────────────────────────────┐
│   collecting and analyzing data │
└─────────────────────────────────┘
              ↓
┌─────────────────────────────────┐
│   summarizing results           │
└─────────────────────────────────┘
```

The symbols on the bottoms of plastic containers indicate their composition and how easily they can be recycled.

Scientists do not always follow all of these steps or approach a problem in a step-by-step fashion, but the basic process is always the same: to try to explain natural phenomena and perform experiments to test these explanations.

The scientific method is illustrated by the following example: A variety of different types of plastic—for example, soft drink bottles—can be recycled to make new products. How would you decide what types of plastic can be recycled into new products?

The following table summarizes the steps that a scientist might go through in order to answer this question.

Step	Questions
considering a problem	How can you tell which samples of plastics can be recycled easily?
forming a hypothesis	Samples that can be remolded by heating are thermoset plastic and can be recycled. Samples that cannot be remolded are thermoplastic and cannot be recycled.
designing an experiment	How could you test a sample to see if it is thermoset or thermoplastic plastic?
performing the experiment	Heat each sample and let it cool.
collecting and analyzing data	Observe what happens to each sample and organize the data to show any patterns that emerge.
summarizing results	Some samples could not be remolded. These are thermoplastic and cannot be recycled.

Try These

1. Using the steps in the scientific method, outline how you would try to answer one of the following questions:

   a. Do sunscreens decrease the effect of ultraviolet radiation on skin?

   b. What effect does deforestation have on world climate?

2. What knowledge is important to the scientists in 1a and 1b?

3. Are 1a and 1b examples of pure or applied science?

1.4 ACTIVITY
SCIENCE AND TECHNOLOGY— PROBLEM-SOLVING PARTNERS

There is a close relationship between science and technology. Sometimes science influences technology and at other times technological advances have posed questions and problems to be solved by scientists. In this activity you will be provided with three case studies. Using the definitions of pure science and applied science (technology), you will come to the realization that scientists and technologists are problem-solving partners.

Method

1. Read each of the following case studies.
2. Do the following for each case study.

   a. State the discoveries made by the pure scientist.

   b. State the application that was made by the applied scientist.

Vulcanized rubber is essential for manufacturing modern tires.

Case Study A: It was a Goodyear

Natural rubber is a useful substance but it is easily affected by temperature extremes, solvents, and abrasion. The ancient Mayans had cured rubber by burying it in sand. The heat of the sun baked the rubber into a form that could be used for shoes and other useful articles. In 1839 Charles Goodyear accidentally discovered that heating rubber latex with sulfur produced a material that was no longer sticky, but was elastic, water repellent, heat and cold resistant, and resilient. This process is called vulcanization. Vulcanized rubber is much more useful than natural rubber and is widely,used in products such as automobile tires.

Case Study B: An Accidental Discovery

In 1964 Dr. Barnett Rosenberg of Michigan State University studied the effects of electricity on bacterial growth. The experiment involved placing platinum electrodes in a bacterial culture. Surprisingly, he discovered that the bacteria had died. When he isolated the bacteria-killing substance, he found that it contained platinum. Since it stopped bacteria from reproducing, Dr. Rosenberg thought the new chemical might stop the growth of cancer cells. This information was used to produce a new drug used in chemotherapy called cisplatin.

Dr. Barnett Rosenberg, discoverer of the drug cisplatin

Case Study C: A Super Discovery

Karl Muller and J. George Bednorz of IBM Research Laboratory in Zurich, Switzerland shared a Nobel prize in 1987 for their discovery of superconducting ceramics. These materials are a mixture of barium, copper, yttrium, and oxygen with the chemical formula $Ba_2Cu_3YO_7$. Why was this such earth-shattering news to the scientific world? A superconductor is a substance that offers almost no resistance to electricity. Resistance to electricity is sometimes desirable, for example, in the coils in a toaster. Resistance causes the wires to give off heat. Resistance is not desirable in the transmission of electricity from power stations to homes and industry nor is resistance desirable in computer chips. If the chips are crowded too closely together, they can generate enough heat to burn out.

The use of superconductors may lead to frictionless trains and other machines.

The discovery of superconductors may lead to many applications within the next few decades. Their use in powerful electromagnets may enable high-speed trains to be suspended above a track without wheels. Use of superconductors in circuits may enable chips to be more closely packed, making even smaller computers possible.

Questions and Conclusions

1. How does the work of a chemist differ from the work of a chemical technologist?
2. "The development of new and improved technologies is very important to the chemical industry." Do you agree with this statement? Explain your answer, citing examples.
3. Where is most of the pure science research carried out in our society? Where is most of the applied science research carried out in our society? What changes if any would you make with this arrangement?

1.5
PHOTOGRAPHY

Vintage photography
equipment

Photography is an excellent example of technology at work. In the days prior to the development of the first camera, people were unable to capture and keep important memories and visual records of events for future reference. People began to work on the problem and eventually cameras were developed.

The technology of photography has rapidly advanced. It is based on the principle that silver compounds are light sensitive. The first photographs were probably made using silver-plated sheets of copper that had been exposed to iodine vapor to produce silver iodide. A later development that led to the process now used involved paper soaked in solutions of silver nitrate and potassium chloride. Early workers in photography found that silver compounds mixed in an emulsion using gelatin produced better photographs. An **emulsion** is a liquid mixture in which tiny droplets of one liquid, usually a fat, are distributed evenly throughout another liquid. (Mayonnaise is an emulsion.)

Before the invention of film as we know it, cameras used thin glass plates with a silver emulsion poured onto them. The emulsion formed a thin film on the plate. Today's films are strips of plastic such as cellulose triacetate coated with various emulsions. The plastic comes in rolls rather than single plates. This means that the camera does not have to be loaded with a new film for each photograph.

The first commercial color film was introduced in 1935 and originally cost more to print than black-and-white film. Today color film processing is the most common process and the least expensive.

The following are the steps in the technological method as applied to photography.

1. **Need:** To record an event for future reference. Is there a faster way to record an event than to have an artist draw a picture?
2. **Research:** Silver compounds are found to be light sensitive.
3. **Idea:** Can silver compounds be used to make shadows and form a picture?
4. **Plan or blueprint:** A camera is designed using the laws of physics and the chemical knowledge that silver compounds are light sensitive.
5. **Prototype:** The camera is constructed and glass plates are coated with silver emulsion.
6. **Test:** The camera is used to take photographs.
7. **Improve:** The film is refined, using an emulsion. Lenses are used to improve the image produced by the camera.

8. **Market:** The product is shown to be useful for the need it was created to meet. Then, in the later stages of development, competing producers show the advantages of their particular product over others.

Try These

1. On what chemical knowledge is the technology of photography based?
2. Name three ways that technology has improved photography.
3. Why is it less expensive to develop color film than black-and-white?

1.6 ACTIVITY
PHOTOGRAPHY

Photography is many things to many people. For some it is a profession, for others it is a hobby, and for others it is both.

There are many different kinds of professional photographers including portrait, news, sports, and wildlife photographers. Amateur photographers fall into almost as many categories, and for many amateurs their hobby becomes a part-time career.

At some time most photographers want to develop and print their own film. Developing black-and-white film is not difficult, as you will discover in this activity.

Materials

developing tank
timer
clothespins (2)
waste film
400 ASA black-and-white film with pictures on it
developer
stop bath
fixer
thermometer
wetting solution

Modern photography
equipment

Method

1. Use the waste film to practice loading the developing tank with your eyes shut.
2. Load the good film into the tank in a completely dark room. When the tank is closed, you can do the rest of the steps in the light.
3. Fill the tank with developer. Start running the timer when you begin pouring in the developer. The length of time for developing will depend on the type of film and the temperature of the solution. Check with your teacher.
4. When the tank is full, tap it sharply to get rid of air bubbles.
5. With the lid on the tank, turn it upside down once each minute to mix the chemicals.
6. When the developing time is up, uncap the tank and pour out the developer. Some developers can be reused. Check with your teacher.
7. Start timing as you begin to add stop bath to the tank. Agitate the tank regularly. Save the stop bath for reuse.
8. Start timing as you begin to add fixer to the tank. Agitate the tank once each minute. Save the fixer for reuse.
9. Take the top off the tank and place it and the reel under running water for 15 to 20 min. The water temperature should be the same as that of the solutions you used.
10. Soak the film on the reel for 5 min in a wetting solution.
11. Remove your film and hang it up with a clothespin to dry. Put a clothespin on the bottom as well to hold the film straight.

Questions and Conclusions

1. Why is your product called a negative?
2. Why must you agitate the chemicals regularly?
3. Why is a certain time given for each chemical step?
4. Why is temperature a factor?
5. Why did you have to remove the air bubbles?
6. How might developing color film be different?
7. Use a library resource center, if necessary, to find out the purpose of each of the following:
   a. developer   c. stop bath
   b. fixer   d. wetting solution

1.7
CONSUMER PRODUCTS

On a daily basis we are affected by the results of technology; we use the products of technology that are produced by companies to meet many needs. The following are a few examples.

Aspirin (acetylsalicylic acid) is a common household drug that is used for minor aches and pains. There are many different brands of this product on the market. Can you name a few? What are some of the commercial claims you have heard about these different products?

Magnesium hydroxide is used in some products to settle upset stomachs. These products are called antacids. There are many different types of antacids on the market. Can you name a few? What are some of the commercial claims that you have heard about this type of product?

In 1933 the first synthetic laundry detergent was introduced. These detergents quickly replaced soaps because they dissolved faster and more effectively, made more suds, and rinsed away quickly. Today there are many different brands of laundry detergents. What are some of the commercial claims you have heard about detergents?

Advertisements on radio, television, and in magazines often claim that a product is new or improved. If it is "new and improved," this means that some new substance has been added to it or removed from it that has changed it in some way. Formulas of successful products are highly guarded secrets. Chemists working for companies are always looking for ways to improve their companies' products.

Many types of consumer products are similar even though they have different brand names. Therefore, in some cases, one small improvement in a product can make a large difference in popularity and sales.

When a company feels it has significantly improved its product, it advertises the fact in a way designed to affect and attract people. Television, radio, newspaper, and magazine advertisements are the methods usually chosen.

1.8 ACTIVITY
TESTING A CONSUMER PRODUCT

A great deal of information on proper nutrition is available. Many people are concerned about getting all the nutrients they need to be healthy. One way to make your diet more complete is by consuming products that contain supplements. Vitamin C is a common dietary supplement. It is a water soluble vitamin that cannot be made in the body. It is also easily destroyed by high temperature and contact with air.

In this activity you will compare the amounts of vitamin C present in different fruit juices.

Materials

10 mL graduated cylinder
test tubes (8)
test tube rack
eyedropper
lemon juice
other fruit juices (5)
starch solution
iodine solution
sodium bicarbonate (sodium hydrogen carbonate)

Method

1. Prepare a data table similar to the following:

Juice	Drops Added to Produce a Color Change	Cost per mL

2. Label your test tubes 1–8.
3. Pour 5 mL of starch solution into a test tube. Add 5 mL of iodine solution. Observe the color of the solution.
4. Add drops of lemon juice to the test tube one by one, shaking the test tube each time, until the blue color disappears. Record the number of drops it took for the solution to change color. Use the lemon juice as your standard of comparison for the other juices.
5. Repeat steps 3 and 4 with the other juices.
6. In test tube 7, mix 5 mL of lemon juice and a pinch of sodium bicarbonate. Repeat steps 3 and 4 with this mixture.

7. In test tube 8, heat 10 mL of lemon juice. Allow it to cool and repeat steps 3 and 4.

8. Your teacher will supply you with the brand name and cost of each of the juices tested. Use this information to complete your data table.

9. Plot a graph comparing drops of juice required to produce a color change and cost per mL of juice.

10. Determine which juice is the "best buy" for vitamin C content and explain why.

Testing fruit juices for vitamin C content

Questions and Conclusions

1. Which fruit juices contained vitamin C? Which did not?

2. After lemon juice, which fruit juice was the best source of vitamin C?

3. What effect did the addition of sodium bicarbonate have on vitamin C? How did heat affect the vitamin C?

4. Is it true that some vegetables contain more vitamin C than fruit? Select two vegetables and test them.

5. What other factors besides the amount of vitamin C may affect the price of the juice?

1.9
CAUTION—
READ THE LABEL

Advances in aviation technology have made international travel widely accessible. With the increase in the number of people traveling to foreign countries, the use of universal symbols has become more and more important. Symbols have become a sort of "universal language." As an informed consumer, you should be able to interpret symbols. The following common warning symbols are used on consumer products.

Figure 1.1   The symbols show the type of hazard a product contains. The frames around the symbol show the degree of that hazard.

In a lab, you will encounter chemicals that require safe handling. The labels on these chemicals give us more information than those on consumer products. Read these labels carefully to see what precautions should be taken.

Try This

1. Examine the labels of products in the kitchen, garage, and bathroom. Look at the symbols and read the directions. What recommendations would you make for the safe handling and storage of each product?

SCIENCE & SOCIETY

Your Role

It may seem that the benefits of technology far outweigh the disadvantages. There are practical problems facing humans that can be solved by new scientific discoveries which will lead to new technological advances.

Many science-related issues that we are faced with today have a chemical basis. Because we live in a technologically advanced country, it is important to familiarize ourselves with the basic principles of chemistry. This knowledge will lead to a better ability to make decisions.

It is important for each of us to become knowledgeable, responsible citizens of our country and our planet. But you don't have to wait until you are an adult to begin—take Marc Kielburger, for example.

On September 28, 1992 Marc was awarded the Ontario Medal of Good Citizenship. At 15, he was the youngest recipient. The award is given to 12 citizens in the province each year for doing useful and good things for other people on a voluntary basis. Marc is a worthy recipient. He has written two books on the environment, worked with the poor in Jamaica, and given talks at many schools about the environment.

Marc has always been interested in maintaining a clean, safe environment. He first got interested in studying cleaning products when he noticed the hazard labels on the many home cleaning products his parents used.

Marc eventually came up with his own list of home cleaners that were safe, were easy to make, could be used for many purposes, and were environmentally friendly. He published these in a book called *Alternative Home Cleaners*. With the help of the H.J. Heinz Company of Canada, 3500 copies of his book were published. He has had requests for copies of his book from as far away as Europe. Marc has now completed his second book, *Lead: Are We Eating It?*

Marc hopes to go on to study science or law but in the meantime he enjoys gym class (his favorite), sports—especially tennis, baseball, and hockey—as well as rock music.

OPPORTUNITY

Comparing Careers

Many jobs that appear in classified ads require high school math and science. Read the following dialogue and answer the questions that follow it.

High school students Reynaldo, Debbie, and Warren are discussing their future careers.

Debbie: I can expect to earn $2295 per month when I graduate from the Petroleum Technician program.

Reynaldo: If I take a Nursing Assistant program, I can expect to earn about two-thirds the starting salary of a Petroleum Technician.

Warren: If I enter a Plumbing Apprentice program, I'll be earning $26 an hour by the time you graduate.

Questions

Assuming that the students go ahead with their plans, answer the following questions:

1. What will be the annual salaries of each of them the year that Debbie and Reynaldo graduate? (Assume a 40 hr work week and 50 weeks in a work year.)
2. Research the annual costs of each student's education.
3. Research what career advances are possible in each career. Do they require further education?

Exercises

Key Words

applied chemistry

applied science

emulsion

nuclear power

prototype

pure science

scientific method

technological method

technology

Review

1. List three aspects of your life that are affected by applied chemistry.
2. List five technological products that you use.
3. Describe three occupations that are involved in designing and manufacturing technological products.
4. What is the difference between pure chemistry and applied chemistry?
5. Describe the relationship between science, technology, and society.
6. Does technology always improve our standard of living? Give examples to support your answer.
7. Give two examples of technology that uses the chemical process.
8. State the steps in the technological method.
9. On what scientific principles is photography based?

Extension

1. Research the development of a building material and write a report on it.
2. Write to a company for information about a chemical product that interests you. Ask about quality control.
3. Compare and contrast the stages in the technological method with the scientific method.
4. Using the steps of the technological method, trace the development of a product such as the television set.
5. Use the technological method to discuss the improvements made to pens (or another product that interests you) over the last 50 years.

CHAPTER 2
THE SEPARATION OF MATTER

This chapter deals with commonly used separation techniques. These techniques are based on the differences in the physical and chemical properties of the substances in a mixture.

The list of ingredients in any commercial product will show that combining substances takes advantage of the properties of substances to produce a useful and needed product.

Products that you may think of as "pure" are often, in reality, mixtures of several substances. For example, "pure" apple juice is a mixture of water, sugars, fibers, and other substances. This highlights the need to be careful in choosing products.

Key Ideas

- Flowcharts are used to develop and interpret instructions.
- Chemists use physical and chemical properties of substances to identify them.
- The method used to separate a mixture depends on the properties of the materials in the mixture.
- Some common methods of separation are flotation, filtration, distillation, recrystallization, and precipitation.

2.1 ACTIVITY
GO WITH THE FLOW

Flowcharts are used to clarify communications in science and many computer-related fields. Flowcharts are used to develop and interpret instructions. You can use flowcharts to give directions to a computer or even a robot. Flowcharts are used in qualitative analysis to organize steps and procedures.

Table 2.1   Flowchart symbols and their meanings

Symbol	Command	Example
	start or stop	used to begin or end
	step or direction–one or more paths in or out	Examine the sample.
	decision or question	Is the element a metal?
→	direction of the action, connects each shape	

Method

1. Read the following example describing how to open a window:

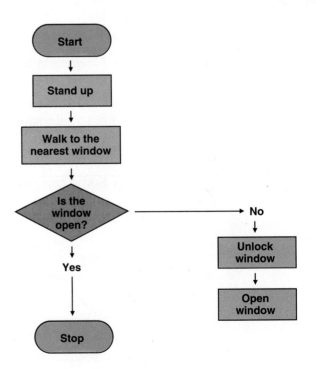

2. Pick three of the following activities and prepare a flowchart for each:

a. hitting a baseball
b. writing with a pen
c. hammering a nail
d. throwing a basketball
e. putting on your shoe
f. brushing your teeth

3. Exchange flowcharts with a partner and try to explain your partner's flowchart.
4. Act out the activities you have charted. Are there any corrections or additions needed?

Questions and Conclusions

1. Did you find it difficult to give clear instructions?
2. What difficulties did you find when preparing your charts?
3. Why are flowcharts useful?

2.2 ANALYZING MATTER SYSTEMATICALLY

Analyzing a situation is often easier if you have a scheme or plan prepared in advance. In Activity 2.1, you analyzed an activity and set up flowcharts to perform the activity. Chemistry is the field of science concerned with the nature and composition of matter and the changes that it undergoes. It is easier to analyze matter if it is grouped into different categories. Figure 2.1 outlines the scheme used by chemists to classify matter. The scheme is based on the characteristics of the substance studied.

Definitions of the Types of Matter

Pure substance: A substance with a fixed composition and constant properties.

Element: A substance that cannot be broken down into simpler substances by chemical means. An element consists of atoms of the same atomic number.

Compound: A substance that is made up of two or more different atoms bonded together in a constant ratio and can be broken down only by chemical means.

Mixture: A mixture consists of two or more kinds of matter, each keeping its own characteristic properties.

Solution: A mixture that is homogeneous. If the solution is a liquid or gas, it is transparent.

Mechanical mixture: A heterogeneous mixture with parts that are visibly distinguishable.

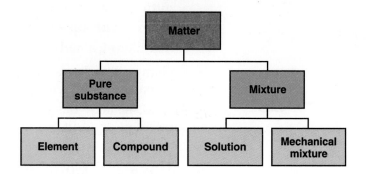

Figure 2.1 The classification of matter

In Activity 2.1, you used flowcharts to help clarify communications. Chemists use flowcharts to help classify and identify materials. When classifying matter, the following flowchart can be used as a guide.

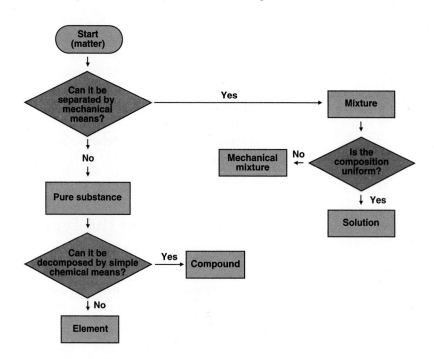

Figure 2.2 Classification flowchart

Try These

1. Use the flowchart to classify
   a. oil and vinegar
      salad dressing
   b. salt
   c. diamond
   d. air
   e. granite rock

2.3 ACTIVITY
CLASSIFICATION
OF MATTER

Chemists often place a substance they are analyzing into a group first to make the identification easier. In this activity you will be grouping or classifying substances into the different groups outlined in Figure 2.1. You will be using chemical composition to assist you in your classification. Matter is first classified into two main groups based on the number of phases (for example, gas, liquid, solid) present. These two groups are homogeneous (consisting of one phase) and heterogeneous (consisting of more than one phase).

Materials

sodium bromide **(NaBr)**(s)
copper sulfate (bluestone) solution **(CuSO$_4$)**(aq)
sulfur and iron **(S, Fe)**(s)
potassium permanganate solution **(KMnO$_4$)**(aq)
water and kerosene **(H$_2$O, C$_{10}$H$_{22}$-C$_{13}$H$_{28}$)**(l)
iodine in alcohol **(I$_2$)**(alc)
oxygen **(O$_2$)**(g)
iron(III) chloride **(FeCl$_3$)**(s)
iron **(Fe)**(s)
water **(H$_2$O)**(l)
air **(N$_2$, O$_2$, CO$_2$, H$_2$, etc.)**(g)
alcohol in water **(C$_2$H$_5$OH)**(aq)

Method

1. Draw a table using the following headings: Name of Material, Chemical Formula, Description.
2. Various materials have been placed around the room. Use Figure 2.2 to classify each sample. Fill in your table.
3. Suggest two different ways you could classify all the materials based on groups with similar characteristics.

Questions and Conclusions

1. Did some of the materials contain more than one substance? If so, which one(s)?
2. Could you use the physical state of the material as a way of classifying it?
3. Is there a difference between a mixture, such as sulfur and iron and a mixture, such as bluestone solution?
4. Draw a flowchart to summarize your results.

2.4
PHYSICAL PROPERTIES

Applied chemists separate the components of a mixture in several ways. Once a substance is separated, it must be identified. You must make sure that the substance you believe you have separated actually is that substance.

Substances can be identified by physical properties such as color, taste, odor, solubility, density, hardness, melting point, and boiling point.

Some physical properties cannot be measured and are called qualitative properties

Table 2.2   Qualitative properties

Qualitative Property	Definition	Terms Used to Describe It
clarity	ability to transmit light	transparent, translucent, opaque
color	color or lack of it	colorless, red, blue, green
odor	characteristic smell	odorless, sweet, sour, burnt, fragrant, choking, nauseating
texture	for solids only; describes feel	crystalline, powder, granular, flaky, wax, fibrous
luster	ability to reflect light	shiny, dull, metallic, greasy, glassy
hardness	ability to scratch other objects	hard, soft, brittle, flexible
viscosity	for liquids only; describes how easily it can be poured	runny, oily, thick, syrupy

Other physical properties can be measured and are called quantitative properties. Every substance has its own unique quantitative properties that can be used to identify it during experimentation.

Table 2.3   Quantitative properties of some substances

Substance	Melting Point (°C)	Boiling Point (°C)	Density (kg/m³)
water	0	100	1 000
gold	1063	2966	19 300
ethanol	−117	78	790
mercury	−39	357	13 600

2.5 ACTIVITY
TWENTY
QUESTIONS

Chemists use the physical and chemical properties of substances to identify them. In Section 2.4, you reviewed some terms used to describe substances. You can use these terms along with the physical and chemical properties listed in a handbook of chemistry to identify unknown substances.

A physical property can be observed without altering the makeup of a substance. Melting point and boiling point are examples of such properties. Chemical properties are those that refer to the ability of a substance to undergo change that alters its structure. Examples of chemical change are iron rusting and milk turning sour.

Materials

test tube	unknown substances (4)
hydrochloric acid	conductivity apparatus

Caution: Hydrochloric acid is corrosive and causes burns. Handle it with care. Wear goggles and an apron. If you spill any on yourself, rinse it off immediately with large amounts of cold water and have your partner tell your teacher.

Method

1. Design a variety of tests to determine the physical and chemical properties of your unknowns.
2. Submit your proposal to your teacher before carrying out any procedure.
3. Draw a data table as shown below.

Number	1	2	3	4
color				
odor				
texture				
hardness				
clarity				
solubility				
reaction with acid				

Questions and Conclusions

1. Which properties were the most difficult to determine?
2. Is there only one characteristic property that you used to identify your unknown?
3. What is your unknown?
4. Use your library resource center to find out the importance of your unknown in industry.

2.6
SEPARATION
OF SUBSTANCES

When a pure substance is needed, impurities must be removed. Applied chemists use many methods for separating mixtures. As you read about each technique, see how many uses you can identify for it. The method chosen to be used in separating a given mixture will often depend on the physical state of the mixture (solid, liquid, or gas).

Flotation

Flotation is used to separate metal ores such as zinc, copper, lead, and nickel from their impurities. Most metal ores occur in the ground mixed together with rock and soil. The ore must be separated from impurities. When the ores are processed, the metallic element can be produced.

The impure ore is crushed and placed in a tank of liquid with water, oil, and detergent. The mixture is stirred and air is bubbled through the tank. The density of the liquid is carefully adjusted so that the soil and rock particles sink to the bottom and the ore particles coated with oil float to the top in the detergent foam and froth. The concentrated ore is skimmed off the top.

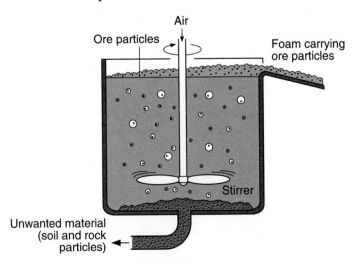

Figure 2.3   Froth flotation technique

Filtration

Filtration is an important method used for separating suspended particles as they pass through a filter, screen, or porous substance. In everyday use, substances are often filtered to remove undesirable particles.

One familiar example of filtration is the purification of water in a swimming pool. Water is continuously pumped through the pool's filters. As the water passes through the filters, dirt and debris that might be harmful is removed.

Distillation

Distillation is another important technique used to separate both solid/liquid and liquid/liquid mixtures. In simple distillation, the liquid is removed from the mixture by evaporation and then re-collected by condensation. An example of this is the condensation of steam on your bathroom mirror after you have showered.

Evaporation and condensation can be caused using the distillation apparatus illustrated in Figure 2.4. When salt water is boiled, the water becomes vapor and flows out of the distillation flask. As it is cooled in another part of the apparatus, it condenses and becomes a liquid again.

When the process is complete, only the solid remains in the distillation flask and the two components of the mixture have been separated. A salt water solution is a homogeneous mixture or solution whose solid is salt and whose liquid is water. Such a solution can be separated by simple distillation.

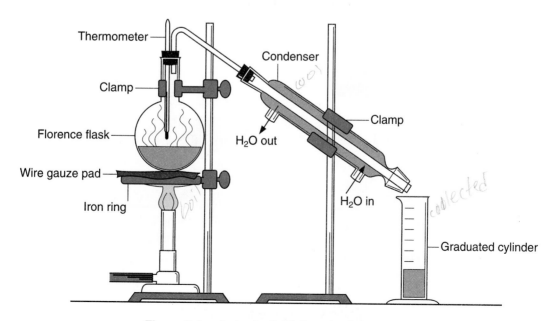

Figure 2.4   A simple distillation of salt water

Salt being raked from
evaporation ponds

Evaporation

Evaporation is used to obtain the solute from a solution or make a solution more concentrated. In some countries salt is removed from sea water by evaporation.

Recrystallization

Some solids dissolve better in hot liquids than in cold. Sugar dissolves easily in hot liquids but not easily in cold liquids. The cooler the solution, the less sugar dissolves. The cooler liquid cannot hold as much of the sugar solution. This behavior of solids in liquids can be used to separate two different substances in an impure mixture of solids. This process is called recrystallization.

Figure 2.5 demonstrates the recrystallization of two substances, **x** and **o**, both of which are soluble in hot water. However, substance **x** will not stay dissolved if the water is too cold. As a result, substance **x** forms a precipitate and falls out of the solution while **o** remains dissolved in the cold water. The next step is to separate **x** from the solution. This can be done easily by filtration. Substance **x** will remain on the filter paper and the filtrate will pass through into the beaker.

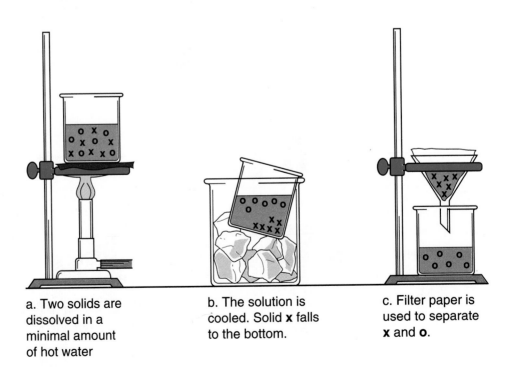

a. Two solids are dissolved in a minimal amount of hot water

b. The solution is cooled. Solid **x** falls to the bottom.

c. Filter paper is used to separate **x** and **o**.

Figure 2.5   Recrystallization as a separation technique

Chromatography

You have probably watched an ink drawing dissolve into a rainbow of colors when water has fallen on it. When an ink drawing comes into contact with water, the water will drag the ink with it as it moves across the paper. Some substances in the ink move more slowly than others. We can apply this property to separate the components in a sample of ink on a document. The word chromatography comes from the Greek words *khroma* and *graphien* and means color writing.

Table 2.4   Methods used to separate some mixtures

Mixture	Method	Chemical Principle	Products
wine	distillation	boiling point	water, alcohol, residue
water-soluble black ink	chromatography	differences in absorption and affinity for solvent	color components of ink
mud/salt	filtration, evaporation	porosity, particle size	water, dirt, salt

The following flowchart illustrates the separation of a mixture of sand, sodium nitrate, and lead chloride.

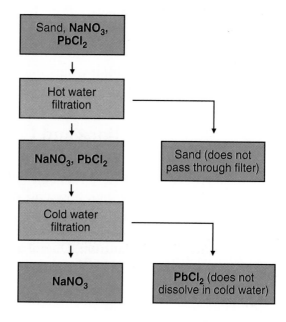

2.7 ACTIVITY
SEPARATION OF A MIXTURE

Some of the methods you will use in this activity are used regularly in industry on a large scale. Other methods are common but are too complex, expensive, or dangerous to carry out in a high school laboratory.

In this activity you will separate different mixtures. The method you choose will depend on the physical state of the mixture and the properties of the components making up the mixture.

> ## Materials
> unknown mixture (3)
> apparatus supplied by your teacher

Separation of a mixture

Method

1. Describe mixture A.
2. Prepare a data table to record your observations with the following headings: Mixture, Components, Physical Properties used in Separation.
3. Use a handbook or chemical dictionary to determine the properties of the components of your mixture, for example, solubility, density, magnetism, particle size.
4. Prepare a flowchart to represent your separation scheme.
5. Have your teacher check your flowchart.
6. Carry out your separation procedure.
7. Submit your final products for evaluation.
8. Repeat steps 1–7 for each of the other 2 mixtures.

Questions and Conclusions

1. What properties did you use to decide on the methods used to separate the components of the mixtures?
2. Explain how you would separate the following mixture: salt, sand, benzoic acid (soluble in water), iodine (insoluble in water but soluble in alcohol), and charcoal. **Caution:** Do not actually perform this separation.

2.8
QUALITATIVE
METHODS

Careful observation is the key to successful chemical detective work. Once a substance is separated, it can be positively identified. You must make sure that the substance you believe you have separated actually is that substance.

In Activity 2.5, you identified substances by physical properties such as color, taste, odor, solubility, melting point, and boiling point.

You can also identify or detect substances through simple tests for the presence of particular ions. An ion is an atom that has gained or lost electrons. If an atom loses electrons, it becomes positively charged and is called a cation. For example, sodium **(Na)** loses one electron to become a sodium cation **(Na$^+$)**. If an atom gains electrons, it becomes negatively charged and is called an anion. For example, chlorine **(Cl)** gains one electron to become a chloride anion **(Cl$^-$)**.

The presence of ions in tap water can be detected by adding test solutions. If the ion is present a solid will form and fall to the bottom of the test tube. This solid is called a precipitate.

The tests for the presence of chloride ions is outlined below.

chloride ion + silver nitrate ⟶ **silver chloride**
(test solution)   (white precipitate)

The following flowchart was used to identify four unknowns in a solution.

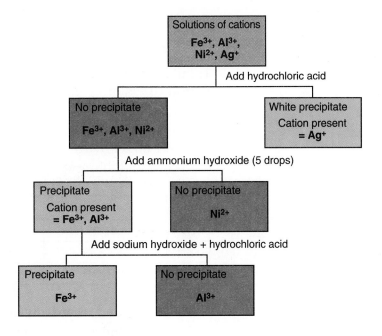

2.9 ACTIVITY
DETECTIVE WORK WITH CATIONS

This activity involves qualitative analysis which is used to identify what substances are present in a sample.

In this activity you will be learning simple tests to detect the presence of three metals: silver (Ag^+), iron (Fe^{3+}), and nickel (Ni^{2+}). First you will test for these ions in a standard solution, which is a solution that is known to contain the particular ion. Then you will test for these ions in an unknown solution. Formation of a blood-red color will confirm that the ion was present.

Materials

test tubes (4)
test tube rack
medicine droppers (4)
pink litmus paper
silver nitrate solution
iron(III) chloride solution

nickel(II) chloride
    solution
unknown solution
hydrochloric acid
ammonia solution
dimethylglyoxime

Caution: Silver nitrate, hydrochloric acid, and ammonia are corrosive and cause burns. Handle them with care. Wear goggles and an apron. If you spill any on yourself, rinse it off immediately with large amounts of cold water and have your partner tell your teacher.

Method

1. Put 2 mL of each standard test solution (silver nitrate, iron(III) chloride, nickel(II) chloride) into separate clean dry test tubes.
2. Label each test tube and describe each solution.
3. Add 1 mL of hydrochloric acid solution to each test tube.
4. Record your observations in a table:

Ion	Test Solution Added	Observation	Conclusion
Ag^+	HCl		
Fe^{3+}	NH_4OH		
Ni^{2+}	NH_4OH		

5. Wait 2 min and record any new observations.
6. To any test tubes that did not form a precipitate, add ammonia drop by drop until the solution turns pink litmus blue. Then add 3 more drops. Record your observations.

7. Add 2 drops of dimethylglyoxime to the test tube that did not form a precipitate. Record your observations. A red precipitate indicates that nickel is present.

8. Repeat steps 1–7 for your unknown. Draw a flowchart to show your results. Use the following flowchart as a guide.

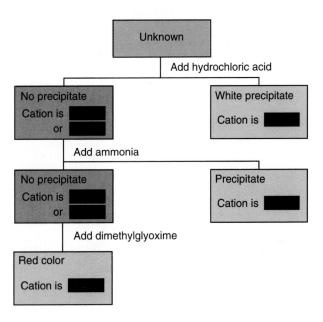

Questions and Conclusions

1. What test solution would you use to test for the following?   a. silver ions   b. iron ions

2. How would you confirm the presence of nickel ions?

3. Distilled water was used to prepare all of the solutions in this activity. Why do you think distilled water was used?

4. Imagine that two bottles have been labeled incorrectly. One of the bottles actually contains an expensive chemical containing silver. What method would you use to find out which bottle contained the silver compound?

5. Imagine that you work for a mining company. You have just been given a sample to test. You do not have the geological reports, but you have been told by the head of the department that either zinc or silver is present. You mistakenly add sodium sulfide to a sample of the test solution and the result is a brownish-blue precipitate. You realize your mistake. What metal is present in this sample? Would you retest other samples to be sure? How would you go about it?

2.10 ACTIVITY
A SALT
PROBLEM

The term "salt" is often misleading. Most foods we ingest contain common salt, which is called sodium chloride. To a chemist, salt is a compound formed when a metal and a non-metal react.

Salts are used in cooking and cleaning. Two salts that are important in the kitchen and laundry room are sodium carbonate (Na_2CO_3) and sodium bicarbonate (sodium hydrogen carbonate, $NaHCO_3$). You will probably be familiar with their common names, washing soda and baking soda.

In this activity you will be identifying which of the two salts is baking soda. The key to the puzzle is in the chemical formula. When baking soda decomposes it produces carbon dioxide gas and water. It is this decomposition into CO_2 and water that makes baking soda so effective at putting out kitchen fires. Washing soda produces carbon dioxide and no water.

Materials

test tubes (3)	washing soda	burner
baking soda	test tube tongs	unknown
limewater		

Method

1. Label 3 test tubes 1, 2, and 3.
2. Place 5 mL of baking soda into test tube 1. Place 5 mL of washing soda into test tube 2 and 5 mL of unknown into test tube 3.
3. Gently heat each test tube and record any changes observed.
4. Place a drop of limewater on a stirring rod. Place the rod at the mouth of test tube 1 and 2 while heating. What did you observe? (A white precipitate in limewater is a positive test for carbon dioxide.)

Questions and Conclusions

1. What happened to the limewater in step 4?
2. What were the products in test tube 1?
3. What were the products in test tube 2?
4. What were the products in the unknown?
5. Was your unknown baking soda or washing soda? Explain your reasoning.

SCIENCE & TECHNOLOGY

Water Treatment Plants

Water treatment plants are used to purify water for our use. The plants use several separation techniques to remove impurities from the water.

Huge settling tanks are used in water treatment plants.

Large particles of impurities such as twigs, pieces of garbage, and even fish are removed as the water passes through a screen. Other relatively large particles of sand and organic matter settle out in reservoirs. At this point, water begins to be treated to help purify it. Alum and lime are added and the white, sticky, jelly-like precipitate that is formed (aluminum hydroxides) traps smaller particles of impurities and carries them to the bottom of a settling tank. The water is drained off the top and then filtered again through sand and gravel. Organic particles, which may cause objectionable tastes and odors, are removed when the water passes through a charcoal filter. The water is usually aerated to increase dissolved oxygen. Finally, the water is treated with chlorine to kill any bacteria which may remain even after filtration. In some areas, sodium fluoride is added to help prevent tooth decay.

CAREER OPPORTUNITY

Water Purification Technician

Water purification technicians work with chemists and chemical engineers at a water filtration plant or a water supply lab, performing a variety of routine tests on water. Collectively, their job is to ensure that the municipalities' drinking water is as pure as possible.

The water samples tested may be taken from the water filtration plant during various stages of the purification process. Some of the tests the technician performs may include tests for turbidity, suspended solids, various ions, organic compounds, and biological contaminants, such as coliform bacteria.

Water technicians use their knowledge of chemistry on a daily basis. Technicians have the satisfaction of knowing that the work they do helps to provide safe, clean drinking water for the entire city.

Exercises

Key Words

chromatography
distillation
evaporation
filtration
flotation
physical property
precipitate
qualitative analysis
qualitative property
quantitative property
recrystallization

Review

1. a. Explain how you would separate the following:
   i.   sodium chloride from sand
   ii. alcohol from water
   iii. salt from water
   iv. ammonium chloride from sodium bicarbonate
   b. Identify the property that allows for the separation of each of the materials in 1a.

2. Give three examples to illustrate applications of filtration in everyday life.

3. Explain the industrial process of flotation.

4. Distinguish between a filtrate and a residue.

5. Define "precipitate" and "precipitation."

6. If a sand and sugar mixture is to be separated by filtration, which substance will be in the filtrate and which will be in the residue?

7. Under what circumstances is precipitation used instead of filtration?

8. Describe the precipitation reaction that occurs when silver nitrate and sodium chloride are mixed.

9. Explain how distillation works. Name any change of state that occurs.

10. What is the difference between crystallization and recrystallization?

Extension

1. Prepare a short report on one application of the distillation process in industry.

2. Use your library resource center to find out how the Dead Sea got its name.

3. a. Name three types of filters used in cars.
   b. In each case in 3a, what is the substance being filtered and what is the residue?

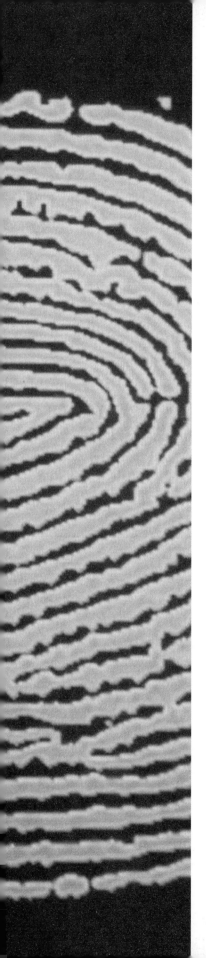

CHAPTER 3
QUALITATIVE
APPLICATIONS

Qualitative analysis is the branch of chemistry responsible for testing for different kinds of materials present in a sample. These substances can be identified through different chemical tests. This chapter will provide you with the opportunity to experience the excitement of searching for unknowns.

The relevance of qualitative analysis will become apparent when performing tests for particular substances. Applied chemistry has its roots in both the chemical and service industries. It is important in the manufacturing of raw materials and in fields such as water purification, toxicology, and forensic chemistry.

Key Ideas

- Qualitative analysis is used to identify what substances are present in a sample.
- Chromatography is a separation technique used in crime detection.
- Many medical and environmental analyses are qualitative tests.

3.1
FORENSIC
METHODS

Forensic science, which is the application of scientific knowledge to the solution of crimes, uses qualitative tests in gathering evidence for criminal investigations, thereby assisting the cause of justice. Homicides and other mysterious deaths, which only a few short years ago were unsolvable, are now often easily solved through the use of technological advances.

When a person is found dead, it is the responsibility of the police to determine the cause of death. This information is supplied by the forensic scientist. Upon examination, it is often possible to determine whether the death was an accident, suicide, or a homicide.

At the scene of a mysterious death, samples of substances, such as paint chips or soil particles, are collected. These can then be compared to samples found on a suspect's car or shoes.

For example, after a hit-and-run accident, paint chips on the accident vehicle can be used to identify the specific car involved in the accident. Modern microscopes and other instruments are used that are so powerful that they can match a paint chip to the paint on the hit-and-run car.

Soil samples also can be used as evidence to indicate that a person was in a particular place. Iron and carbonate ions frequently are present in soil samples and can be detected through the use of chemical tests.

The science of fingerprinting is called dactyloscopy. It is perhaps the best means for identification of humans. Fingerprints found at the scene of a crime can be used to help convict a suspect; however, 12 to 16 points on the print must match those prints that were found at the scene of the crime.

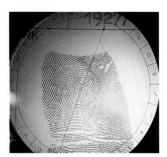

Fingerprint analysis is widely used to solve crimes.

3.2 ACTIVITY
FINGERPRINTING
IN CRIME
DETECTION

No two people in the world, not even identical twins, have the same fingerprints. The ridges which cause fingerprints were first discovered by Malpighi, an Italian anatomy professor, in the late 1600s. The ridges are not limited to the fingers. Babies' footprints are taken after birth for identification. Some pet owners have gone to the extent of having the nose prints of their animals taken in case they should become lost. Fingerprints show up easily on soft surfaces. Since not all surfaces are soft, many different techniques have been devised to obtain a print. In this activity you will be using two techniques for developing a fingerprint.

In Case A, a mystery print was left at the scene of a crime and you will be developing your classmates' prints to see if there are any suspects among you. Fats, fatty acids, and sodium chloride are three of the many chemicals found in sweat. When a print is made, fats and fatty acids are left behind. Fats are only present in the ridges, and when they react with iodine they leave an outline.

In Case B, you are faced with the task of simulating a crime where the criminals have wiped blood from their hands and left faint prints on everything they have touched.

Materials

white paper
iodine crystals
250 mL Erlenmeyer flask
hot plate
tongs
scissors
transparent tape
microscope slide
100 mL beakers (3)
hand lens
tweezers
oven
paper towel
stewing beef
Solution I (acetic acid, naphthalene black, methanol)
Solution II (acetic acid, methanol)
Solution III (acetic acid, water)

Method

Case A: Making fingerprints visible with iodine

1. Cut out a narrow strip of white paper 20 cm long and 3 cm wide.
2. Use your thumb to press down firmly on the end of the paper strip.

Caution: Step 3 must be performed in a fume hood. Do not breathe iodine vapor: it is toxic.

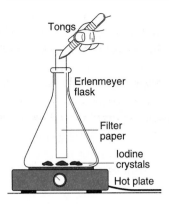

Figure 3.1
Making fingerprints visible with iodine

3. In a fume hood place a few iodine crystals into a flask. Heat strongly with a hot plate until the iodine forms a vapor.
4. Place the strip of paper with the fingerprint into the flask with tongs.
5. Remove the strip with tongs once the print has become visible and immediately cover with transparent tape.
6. Compare the fingerprint with the mystery print at the front of the room. Does it match?

Case B: Developing fingerprints left in blood

1. Rinse and dry your hands.
2. Press your index finger into a piece of uncooked stewing beef.
3. Press the same finger onto the end of a microscope slide.
4. Place the slide in a drying oven for 10 min.
5. Use tweezers to remove the slide from the oven.
6. Place the slide into a 100 mL beaker as shown in the margin.
7. Pour 25 mL of Solution I into the beaker to cover up the print.
8. Gently swirl the solution round in the beaker until the fingerprint is clearer.
9. When the fingerprint is clear, use the tweezers to remove the slide and place in a clean 100 mL beaker.
10. Pour 25 mL of Solution II into the beaker, swirl and let sit for 2 min.
11. Remove the slide with tweezers and place in a 100 mL beaker.
12. Pour 25 mL of Solution III into the beaker and swirl for 5 min.
13. Carefully rinse the slide and pat dry with a paper towel.
14. Examine your fingerprint with a hand lens.
15. Draw a picture of your fingerprint.
16. Compare your print with those of your classmates.
17. Wash your hands thoroughly after this activity.

Figure 3.2
Making fingerprints visible with solutions

Questions and Conclusions

1. What effect would washing your hands before starting Case A have on your print?
2. Why did you cover the print with transparent tape?
3. List the circumstances where each of these methods of fingerprinting would be suitable.

SCIENCE & TECHNOLOGY

Color Changes and Diagnostic Tests

When a color change occurs as a result of a chemical reaction, it is often easy to see. If the reaction can be applied to a medical situation or an environmental condition of importance, then it can be used to diagnose or analyze the situation.

Many diagnostic medical tests rely on color changes. For example, people with diabetes have to be very aware of their blood sugar level. They can determine this information for themselves by testing their urine for the amount of sugar present. This is done using strips of paper and matching the color to a color-coded chart.

Color tests can also be used to determine pregnancy and blood alcohol levels.

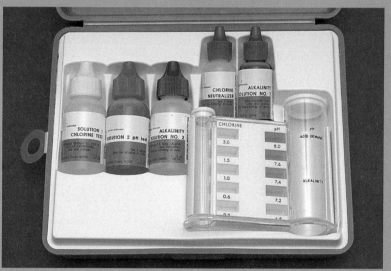

"Backyard chemists" often use color in swimming pool chemistry. They determine the chlorine and pH levels in pool water by comparison to a color chart.

3.3 ACTIVITY
IDENTIFYING
SOIL

The type of rock and rotted plants which make up soil give it its characteristics. Rocks are made of particles, which are made up of chemicals. The distinct red and yellow colors in soil are normally attributed to iron compounds. The acidity of the soil is caused by the accumulation of acids, both organic and inorganic. Soils can be alkaline when there is a high concentration of carbonates. Police use the color of soil as evidence. If traces of soil are found on the shoes of the individual, the soil can be examined to determine whether the person was present at the scene of the crime.

The distinctive white sand of Florida (left) and red soil of Prince Edward Island (right) are a result of chemicals in the soil.

In this activity universal indicator will be used to determine the acidity of the soil. Figure 3.3 indicates the scale of acidity (**pH scale**). You will be testing a sample that has been removed from the shoe of an individual suspected of being at the scene of a crime (A). You will compare it to the sample taken from the scene of the crime (B).

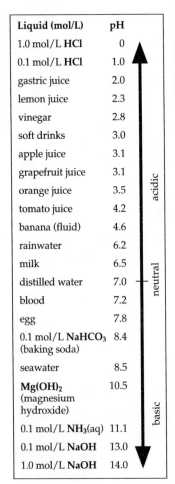

Liquid (mol/L)	pH
1.0 mol/L **HCl**	0
0.1 mol/L **HCl**	1.0
gastric juice	2.0
lemon juice	2.3
vinegar	2.8
soft drinks	3.0
apple juice	3.1
grapefruit juice	3.1
orange juice	3.5
tomato juice	4.2
banana (fluid)	4.6
rainwater	6.2
milk	6.5
distilled water	7.0
blood	7.2
egg	7.8
0.1 mol/L **NaHCO₃** (baking soda)	8.4
seawater	8.5
Mg(OH)₂ (magnesium hydroxide)	10.5
0.1 mol/L **NH₃(aq)**	11.1
0.1 mol/L **NaOH**	13.0
1.0 mol/L **NaOH**	14.0

acidic

neutral

basic

Figure 3.3
The scale of acidity (pH)

Materials

test tubes (6)	watch glasses (2)
test tube rack	distilled water
retort stand	2 mL 1 M **HCl**
test tube clamp	2 mL 1 M **NaOH**
filter	scoopula
filter paper	soil samples (2) (Exhibits A and B)

Caution: Hydrochloric acid and sodium hydroxide are corrosive and cause burns. Handle them with care. Wear goggles and an apron. If you spill any on yourself, rinse it off immediately with large amounts of cold water and have your partner tell your teacher.

Method

1. Label one test tube Exhibit A.
2. Set up a data table similar to the following:

Sample	Universal Indicator Color	Acid/Neutral/Base
Exhibit A		
Exhibit B		
Acid		
Base		
Water		

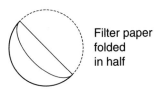
Filter paper folded in half

Filter paper folded in quarters

Filter paper ready for funnel

Filter paper in funnel

Figure 3.4
Folding filter paper

3. Obtain two scoops of Exhibit A on a watch glass.
4. Record the color of the soil sample in your data table.
5. Using a scoopula transfer the sample into the test tube labeled Exhibit A. Add 20 mL of distilled water.
6. Cork the test tube and shake.
7. Fold the filter paper as illustrated in the margin.
8. Place the filter paper in the filter and pour Exhibit A into the filter paper. Collect the liquid in a test tube.
9. Dip a piece of universal indicator paper into the liquid filtered in step 8 and record the color of the paper.
10. Complete the data table for Exhibit A.
11. Repeat steps 1–10 for Exhibit B.
12. Pour 2 mL of acid solution into test tube and dip a strip of universal indicator paper into it. Record the color.
13. Repeat step 12 with 2 mL of base and then with 2 mL of water.

Question and Conclusions

1. What color did universal indicator turn in acid? water? a base?
2. Was Exhibit A acidic, basic, or neutral? Explain.
3. Was the suspect at the scene of the crime? Explain your reasons for coming to your conclusion.

3.4 ACTIVITY
FORGERY

Forensic scientists may be called upon to determine whether a document has been altered. If the words have been changed or replaced by another person, it is referred to as forgery. Forgery can be detected by examining the ink present in the original document and comparing it to the suspected forgery. In this activity you will be using a method called chromatography to determine whether a document is a forgery.

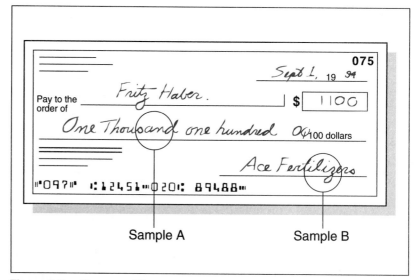

Figure 3.5   Document X

Materials

Ink sample A
Ink sample B
250 mL beaker

straw
rectangular filter paper

Method

1. Obtain a piece of filter paper and label it as shown in Figure 3.6.
2. Using a capillary tube drop a sample of A on the filter paper 2 cm from the bottom edge.
3. Place the top of the filter paper through a slit in a straw.
4. Carefully pour enough water into the beaker to just cover the bottom of the beaker.
5. Place the paper in the water, but be careful that the water does not touch the sample.
6. Allow the paper to sit until the water has moved up to the straw.

Figure 3.6
Chromatography apparatus

7. Remove the filter paper and allow it to dry.
8. Repeat steps 1–7 with Sample B.
9. Draw diagrams of both Sample A and Sample B.

Questions and Conclusions

1. How did Sample A and B compare?
2. How did Sample A and B differ?
3. Was the document forged? State evidence for or against.

3.5

QUALITATIVE ANALYSIS IN MEDICINE

Dr. Frederick Banting (1891–1941) (right) and Dr. Charles Best (1899–1978) (left) in 1921 with the first dog to be kept alive with insulin

Medical research and its related technologies have vastly improved the human condition. Many medical analyses are qualitative tests for substances that may be present in the blood or urine. Many diseases that were killers in the recent past are either extinct or are controllable by modern-day medicines.

An example of this is insulin, which was discovered by Dr. Frederick Banting and Dr. Charles Best at the University of Toronto in 1922. Insulin, which is injected as a treatment for diabetes, increases the life expectancy of diabetics. Diabetes is the condition which results when there is a lack of the hormone called insulin in the blood. This hormone is needed to break down sugars and fats in the body; therefore, diabetic people have an abnormally high sugar level in their blood. Since the kidneys do not normally remove sugar from the blood stream, this sugar is passed from the body in urine. The urine of an untreated diabetic person is frequently quite acidic and has a high level of sugar. Untreated diabetes causes serious illness and death.

Dr. Banting and Dr. Best isolated the hormone insulin from the pancreases of dogs. When this hormone is injected into diabetic people, their blood sugar is reduced to normal levels.

Try This

1. What have been some recent advances made in medicine in the past year which have made newspaper headlines?

3.6 ACTIVITY
URINALYSIS

Urine has often been used to evaluate a person's health. More than 5000 years ago in Egypt, diabetes was diagnosed by tasting the urine for sweetness. It wasn't until the invention of medical technologies that urine became an accurate clue to bodily function. We excrete approximately 1500 mL of water every day. Normal urine is about 95% water and 5% dissolved salts, organic wastes, and hormones. Substances such as albumin, glucose, blood, pus, and bacteria indicate a health problem.

In this activity you will test urine samples for the presence of a substance not normally found in urine. The presence of sugar in urine may indicate *diabetes mellitus* (literally, honey diabetes); the presence of protein or a high pH may indicate an infection of the urinary system.

Materials

10 mL graduated cylinder
hot water bath
test tubes (6)
pH paper strips
disposable vinyl gloves
safety glasses
Benedict's solution
Biuret solution
artificial urine samples (3)
unknown artificial urine samples

Caution: Wear goggles and gloves during this activity. Benedict's solution is poisonous when ingested. While heating solutions, always point the open end of the test tube away from yourself and others.

Method

1. Label your test tubes 1: pH, 2: sugar, 3: protein, 4: unknown, 5: Benedict's solution, 6: Biuret solution.
2. Pour 2 mL of each urine sample into separate test tubes 1, 2, 3 and 4 mL of unknown urine sample into test tube 4.
3. Find the pH of test tube 1 by dipping a strip of pH paper into it. Record the pH.
4. Add 3 mL of Benedict's solution to test tube 2 and place it in a hot water bath for 5 min. Record any color change.

5. Add 3 mL of Biuret solution to test tube 3. Record any color change.
6. Find the pH of tube 4 and divide its contents into test tubes 5 and 6.
7. Add 3 mL of Benedict's solution to test tubes 5. Put the tube in a hot water bath, and record any color change.
8. Pour 3 mL of Biuret solution into test tube 6. Record any color change.

Questions for Discussion

1. Which samples contained sugar? protein?
2. What is the significance of sugar in the urine? of protein?
3. What might a high pH in a urine sample indicate?

Extension

1. Visit a pharmacy and compile a list of names of test strips used for different purposes.

3.7
WATER POLLUTION

Water is one of our most important and plentiful resources. It is essential to all forms of life. We use it for drinking, recreation, transportation, cooking, and cleaning. We take food from rivers, lakes, and oceans. Yet humans, in our attempts to enjoy a better life, are damaging one of the world's essential resources.

Water pollution is not a new problem. It has been with us since the beginning of humankind. The problems associated with it became obvious only when large groups of people began living together and using the same water supply for their waste disposal.

It was a generally accepted practice in Europe by 1880 to use water to flush away human waste. As a result, many water supplies became unfit for human consumption. Unaware of this, people continued to drink the contaminated water.

Very little was known at that time about the chemistry of the human body. No one knew, for example, that disease-causing bacteria living in the human intestine and present in human waste could cause infection and death if reintroduced into the body by drinking water contaminated with it.

Infection with the bacterium *Vibrio cholerae* causes cholera, a highly contagious disease that killed some 20 000 people in London, England between 1849 and 1853. This disease was so infectious that it quickly spread throughout Europe, causing many deaths. Contamination of water by infectious bacteria and parasites is still a serious problem in much of Asia, Africa, and South America.

In Europe and North America, public water supplies are generally safe to drink. However, many bodies of water are polluted by the chemical byproducts of industries, such as power supply companies, pulp and paper mills, mines, and chemical product industries.

3.8 ACTIVITY
DETECTING PHOSPHATES IN THE KITCHEN

Phosphorus is a mineral which is essential for healthy growth in both animals and plants. The most common form that phosphorus is found in is the phosphate ion. Phosphates can be found in food, canned beverages, fertilizers, and cleansers. The form they are found in depends on their use.

Table 3.1   Common phosphorus compounds

Name	Chemical Formula	Use
monosodium phosphate (sodium hydrogen phosphate)	$NaHPO_4$	controls acidity in food
sodium triphosphate	$Na_5P_3O_{10}$	detergents (breaks up dirt)
calcium dihydrogen phosphate	$Ca(H_2PO_4)_2$	fertilizers, baking powder

Materials

samples (food, beverage, detergent, fertilizer)
sodium phosphate
distilled water
red litmus paper
hydrochloric acid
phosphate test solution
test tubes (5)
eyedroppers (5)

Caution: Hydrochloric acid is corrosive and causes burns. Handle it with care. Wear goggles and an apron. If you spill any on yourself, rinse it off immediately with large amounts of cold water and have your partner tell your teacher.

Method

1. With an eyedropper, place 0.6 g of sodium phosphate in a test tube. Add 20 mL of distilled water. Shake the tube until all of the sodium phosphate has dissolved. (Add more water if necessary.)
2. Dip a piece of red litmus paper into the solution. Add hydrochloric acid one drop at a time until the paper changes color.
3. Add a dropperful of the phosphate test solution and mix.
4. Observe what happens. A white precipitate indicates that a phosphate is present.
5. Repeat steps 2–4 for each sample. Record your results.

Questions and Conclusions

1. Which samples contained phosphates? How did the amounts of phosphates compare in each sample?
2. Examine the labels of each of the phosphate samples. List the form of the phosphate.
3. Find three other products that contain phosphates.

Extension

1. In the 1960s the use of high-phosphorus detergents resulted in the eutrophication of lakes. What is eutrophication? How was this problem solved?

CAREER OPPORTUNITY

Forensic Laboratory Technician

Technicians in a forensic laboratory examine and analyze objects for the police and other investigative branches of the government. Their work is done in many different areas ranging from the identification of firearms to voice analysis.

Technicians in the chemistry branch of the forensic laboratory analyze soils, paint, glass, metals, explosives, and plastics using infrared spectroscopy and gas chromatography.

Gas chromatography can also be used to separate and identify the components of a mixture such as blood, paint, or clothing fibers.

The results of the technician's and laboratory's findings are often used as conclusive evidence in courts of law. This makes forensic science a highly effective tool in solving crimes and in greatly increasing public safety.

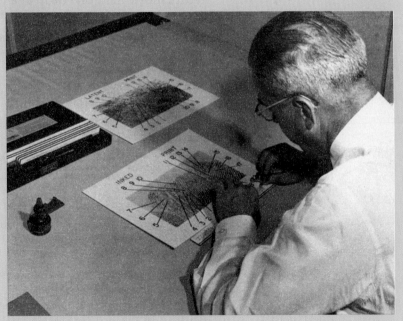

A police fingerprint expert marks points of identification for use in court testimony.

Exercises

Key Words

diabetes

forensic science

insulin

pH scale

Review

1. What is the role of the forensic scientist?
2. What two methods did you use to make fingerprints visible? Explain how they worked.
3. Why are fingerprints used as evidence in solving a crime?
4. What element is responsible for the red and yellow color observed in soils?
5. Why are some soils acidic? basic?
6. What instrument is used to measure pH?
7. What is a positive test for
   a. carbonates?   b. protein?   c. sugar?
8. What hormone is lacking in an individual who has been diagnosed as diabetic?
9. What condition is indicated if the following was detected in urine?
   a. high pH   b. protein   c. sugar
10. How could large groups of people cause a water supply to become contaminated?
11. Describe a positive test for phosphate.
12. What would a reading of pH 3 indicate about the water?

Extension

1. Fingerprinting has been used in some cases to convict an individual. Use your library resource center to research such cases in the past year.
2. Obtain a sample of water from a source near your home or school. Carry out an analysis using chemical test kits. If there is a pollutant, trace its source and write a letter to the polluters.
3. Describe how you would make your own fire extinguisher. If you wish to build it, show your plans to your teacher first.

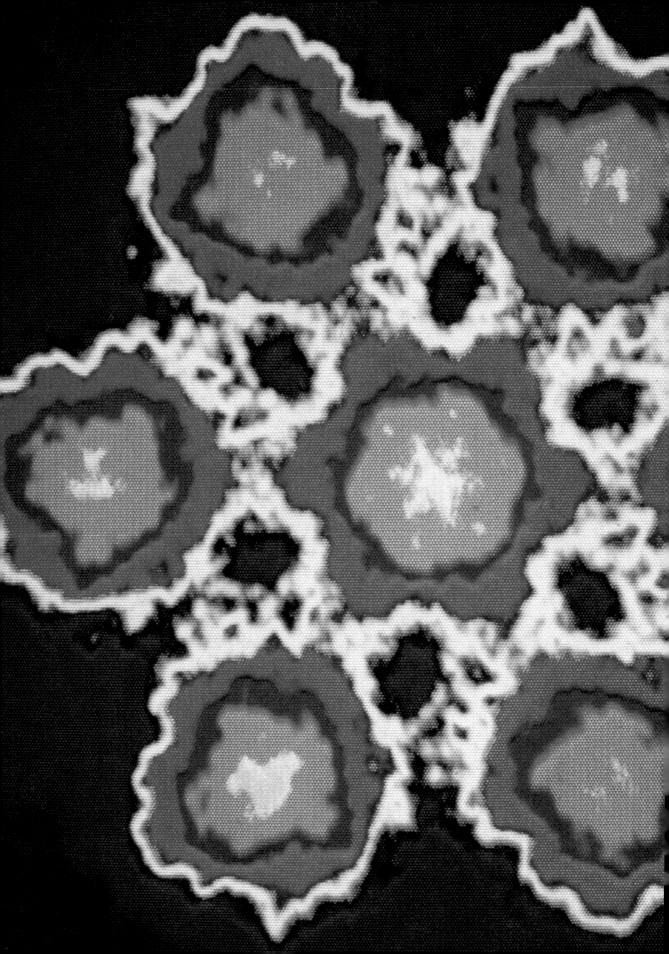

CHAPTER 4
EARLY ATOMIC
THEORIES

Almost all matter is composed of tiny particles called atoms. In this chapter, you will learn about the theory of the atom and how this theory developed through history. Many of the important experiments on the structure of the atom were done near the beginning of the 1900s.

The discovery of subatomic particles that make up every atom has had a tremendous impact on our lives. As you work through this chapter, you will see that many of the recent technological developments designed to improve our lives make use of our knowledge of the structure of atoms.

Key Ideas

- Models are used to show how something that cannot be seen works.

- Our present model of the atom is the result of the cooperation of scientists from many countries over a long period of time.

- Advances based on knowledge of subatomic particles have greatly improved the quality of our lives. This technology includes television, computers, radar, X rays, and radiation treatment of cancer.

- Elements can be broken down into atoms, which in turn can be broken down into protons, neutrons, and electrons.

An electron micrograph of uranium atoms within a crystal. Each purple circle is a single uranium atom. The colors were added artificially to enhance the image. The atoms have been magnified approximately 100 million (10^8) times.

Air Traffic Controller

Air traffic controllers perform an essential role at every major airport in the world. The air traffic controller is responsible for keeping track of aircraft as they travel through a certain controlled airspace. The controller uses a radar screen (cathode ray tube) to monitor the distance, speed, and direction of all aircraft in the airspace. They watch the radar screen to ensure that the aircraft are at least 8 km away from each other or separated by 300 m in altitude.

The controller must be in contact with the pilot/navigator of the plane to give instructions about the altitude the aircraft should be at, and the direction and speed it should travel. As an aircraft moves out of the controller's airspace, the controller must communicate by telephone with a controller in the area that the aircraft is moving into and pass responsibility for the aircraft along to the next controller.

Air traffic controllers must have a good knowledge of physics. They are constantly dealing with distance, time, speed, and the relationships among them. They must also be knowledgeable of the different types of aircraft and their capabilities. Being an air traffic controller is considered a stressful but rewarding career.

Candidates for air traffic control must first pass an aptitude test and an interview. Once a candidate is accepted, the training program lasts about two years.

Exercises

Key Words

atom

black box

cathode ray

Dalton's Atomic Theory

electron

elementary particle

fluorescence

model

neutron

nuclear atom

nucleus

particle theory

postulate

proton

solar system model

subatomic particle

X ray

Review

1. State Democritus's theory in your own words. *Democritus*
2. Who first used the term "atom"? *Democritus*
3. a. What contributions did Boyle make to modern science?
   b. How does today's definition of an element differ from Boyle's definition?
4. a. Define the term "black box."
   b. Name four things that are black boxes to you.
   c. For each answer to 4b, name a person or group of people for whom they are not black boxes.
5. a. State the postulates of Dalton's Atomic Theory.
   b. Which parts of Dalton's Atomic Theory are no longer accepted? Explain why.
6. Which subatomic particles make up nearly all of the mass of an atom? *electron*
7. Compare the mass of a proton to the mass of a neutron.
8. Explain the difference between
   a. a charged object and a neutral object
   b. a positively charged object and a negatively charged object
9. Which of the three main subatomic particles can be removed from an atom by friction?
10. a. Why did Thomson believe that all atoms contain electrons?
    b. Why did Thomson also believe that atoms contain positive charges?
11. What is the charge on an alpha particle? a beta particle?
12. What two particles make up the nucleus of an atom?
13. Explain why, in Rutherford's alpha-scattering experiment, most of the alpha particles traveled straight through the foil. Why were some of the particles deflected?
14. Describe Rutherford's model of the atom based on his experimental evidence.

CHAPTER 5
MODERN ATOMIC THEORY

From the ancient Greeks to Ernest Rutherford, much was learned about the structure and behavior of atoms. In the early 1900s, Neils Bohr made a startling discovery that drastically changed atomic theory.

In this chapter you will learn more about subatomic particles, their arrangement in atoms, and the energy related to them.

Key Ideas

• Electrons exist in energy levels at varying distances from the nucleus.

• Electrons can absorb energy and move to a higher energy level.

• Electrons can lose excess energy and return to a lower energy level.

• Bohr-Rutherford diagrams can be used to represent the nucleus of an element and the electrons in their energy levels.

The brilliant colors of fireworks are caused by the excitation of atoms in metal salts.

5.1 ACTIVITY
SPECTROSCOPES AND GAS TUBES

Historically, experiments just like the one you are about to do led to the quantum model of the atom proposed by Niels Bohr in 1913.

You know that white light can be broken down into its component colors by a triangular prism. The spectroscopes you use in this activity will have the same effect.

> ## Materials
>
> hand-held spectroscopes or large spectroscope on a stand
> gas tubes filled with hydrogen, helium, neon, sodium
>   vapor
> power supply
> colored pencils

Hold the spectroscope about 1 m away from the gas tube.

Method

1. Use the spectroscope to examine a source of white light such as the lights in your classroom.
2. The diffraction gratings in the hand-held spectroscopes may need to be adjusted. Adjust them by rotating the plastic end of the spectroscope, not the slit end.
3. Keep the slit parallel to the light source and slightly to one side of the light source. Do not look directly at the light source as you look through the eyepiece.

4. Look inside the spectroscope to one side of the slit or the other. Turn the eyepiece until you see a spectrum.

5. Your teacher will put gas tubes in the power supply and turn it on for you. Use the spectroscopes to examine the spectrum emitted by each gas. Look for any fine (very narrow) lines in the spectrum.

6. Use colored pencils to draw the line spectrum for each gas.

Questions and Conclusions

1. What type of spectrum did you see when you looked at the white light source through the spectroscope?

2. How many lines did you see when you looked at the hydrogen gas tube? What colors were the lines?

3. What are the two kinds of spectra observed in this activity?

4. Name two ways that this type of scientific knowledge is commonly used in today's technology.

5.2
THE BOHR ATOM

Niels Bohr (1885–1962)

The solar system model of the atom seemed to explain all of the experimental evidence that was available at the time. However, it raised some interesting questions.

Rutherford showed that the nucleus contained all of the positive charge and most of the mass of the atom. Electrons orbited the nucleus and were negatively charged. If this were true, why weren't the electrons pulled into the nucleus by the forces of electrical attraction? In other words, why didn't the atom collapse?

The answer to this problem was suggested in 1913 by Niels Bohr, a Danish physicist, while he was experimenting on samples of hydrogen gas. He found that the hydrogen gas spectrum was made up of different colors. The four colors he observed were red, bluish-green, indigo, and violet.

But what did this mean to Bohr? Bohr knew that light is a form of energy. He also knew that white light is composed of all the colors in the visible spectrum. He reasoned that the lines he saw represented different amounts of energy. Bohr suggested that each line represented a different energy level or position that the electron in a hydrogen atom could occupy. This type of spectrum is called a line spectrum. Each different type of atom has its own characteristic line spectrum.

SCIENCE & TECHNOLOGY

Chemical Composition and Age of the Stars

A star is a hot mass of gases which gives off radiant energy. Most stars shine for billions of years. The energy emitted by stars is produced by a nuclear fusion reaction that takes place in the core of the star. The hydrogen nuclei are fusing to form helium nuclei and massive amounts of energy.

By using spectroscopy, scientists can determine the mixture of gases present in a star. Analysis of starlight with a spectroscope is used to determine the kinds and amounts of the materials present in the star. Astronomers know the rate at which the hydrogen in a star is converted to helium, carbon, and finally, iron. The composition of the star changes over time. It can therefore be used to determine the age of the star.

Figure 5.1
These spectra were drawn on a computer and are similar to the spectra of some well-known stars. The black lines indicate that light has been absorbed by gases in the stars' outer layers. Most of the vertical black lines are caused by hydrogen, but some indicate the presence of calcium or sodium.

Black horizontal bands separate different stars' spectra. As you can see, each star has a different composition. For instance, Star A's spectrum is different from Star H's.

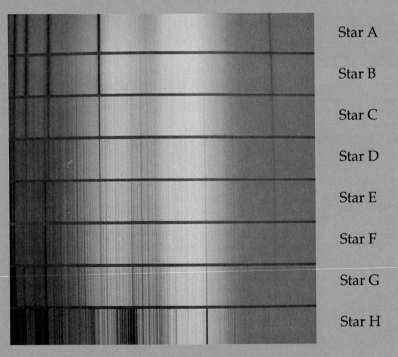

Star A

Star B

Star C

Star D

Star E

Star F

Star G

Star H

Stars are born, they live, and they die. Stars are formed from clumps of interstellar gases and dust. Very large and massive stars consume their energy more rapidly than smaller stars.

A star dies when its supply of nuclear energy is used up.

5.3
THE ENERGY-LEVEL MODEL

The energy-level model assumes that electrons can orbit the nucleus only at certain fixed distances from the nucleus. These are called energy levels. The energy level of an electron depends on its distance from the nucleus. The closer the electron is to the nucleus, the less energy it has. The farther the electron is from the nucleus, the more energy it has.

This model proposes that there are many imaginary spheres of increasing radius, with the nucleus at the center. These spheres represent different energy levels.

Figure 5.2   A representation of the energy-level model, showing the levels that electrons can occupy

5.4
ENERGY-LEVEL CHANGES

Electrons can jump from one energy level to another. We can compare this jump to throwing a ball up a staircase. Energy is added to the ball when it is thrown. The steps in the staircase represent energy levels for that particular atom.

The ball will land on a step or energy level and begin to bounce back toward you down the steps or energy levels. With every bounce it loses or gives off a certain quantum, or amount of energy. It will continue to bounce down the steps, releasing energy with each successive bounce until it reaches its original energy level.

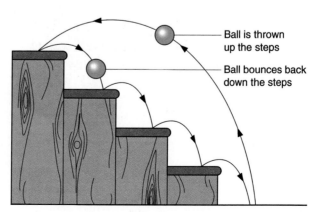

Ball is thrown up the steps

Ball bounces back down the steps

Figure 5.3   Another representation of the energy-level model

Just as the ball on the steps goes to a higher level when energy has been added to it, an electron moves to a higher energy level when energy is added to it. An electron that has energy added to it is in an excited state, and it will jump to a higher energy level farther from the nucleus. When the energy source is removed it will fall back to its original energy level, or ground state, giving off light of a characteristic color for that particular type of atom.

The ball may fall two or three steps at a time, but just as the ball cannot possibly fall halfway between steps, the electron cannot fall halfway between energy levels. The line spectrum gives us evidence that the electrons fall from one energy level to another. In so doing, they give off definite amounts of energy.

5.5 ACTIVITY
FLAME TESTS

When an atom is heated to a high temperature, one or more of its electrons are raised to a higher energy level. Later, the electron or electrons return to a lower energy level and give off the excess energy in the form of visible light of a characteristic color.

In the following flame tests, you will observe the flame colors of various metallic compounds, both with your naked eye and through a spectroscope. In this activity follow either method A or B, as instructed by your teacher.

What element is being flame-tested here?

Materials

burner

nichrome test wires with insulated handles or T-pins in wooden handles

spectroscope

concentrated hydrochloric acid

lithium chloride

potassium carbonate

sodium chloride

strontium chloride

copper(II) chloride

copper(II) sulfate

potassium chloride

strontium nitrate

sodium bicarbonate (sodium hydrogen carbonate)

selected unknowns (2)

Caution: Concentrated hydrochloric acid is corrosive and causes burns. Handle it with care. Wear goggles and an apron. If you spill any on yourself, rinse it off immediately with large amounts of cold water and have your partner tell your teacher.

Method

A. Using Nichrome Test Wires

 1. Prepare a chart as shown below.

Substance	Flame Color
lithium chloride	
sodium chloride	
potassium chloride	
strontium chloride	
copper(II) chloride	
copper(II) sulfate	
strontium nitrate	
sodium bicarbonate	
potassium carbonate	
unknown 1	
unknown 2	

2. Light the burner and adjust it so that the flame is almost colorless. Hold the nichrome test wire in the flame momentarily.
3. Clean the nichrome wire by dipping the loop into concentrated hydrochloric acid and returning it to the flame. Repeat this process until the nichrome wire no longer gives any color to the flame.
4. Dip the tip of the nichrome test wire into the lithium chloride solution and then hold the wire in the flame. View this flame color with your naked eye and through the spectroscope.
5. Repeat step 3 to clean the wire.
6. Repeat step 4 for each of the other solutions. Clean the wire each time.
7. Repeat step 4 for the two unknowns provided. Clean the wire between tests.

Method

B. Using T-pins

1. Follow the same procedure as in method A except place solid crystals of each of the compounds on the T-pin by dipping the T-pin into water to wet it, then into the various solids to be tested.

Questions and Conclusions

1. Make lists of the substances that produced the same flame color.
2. What is common to substances with flames of the same color?
3. What is a positive flame test for each of the following?
   a. copper   c. lithium   e. sodium
   b. potassium   d. strontium
4. What metallic element is present in Unknown 1? Unknown 2?
5. A reddish brown rock was held in a very hot burner flame. The flame appeared emerald green in color. What metal was most likely present in the rock?

MINI-ACTIVITY
FIREWORKS

Fireworks make use of the same phenomenon that occurs in flame tests. The atoms of metallic elements absorb some additional energy. The energy raises electrons to higher energy levels. This energy is then released as the characteristic color or spectrum for the element. For example, the bright red color in fireworks displays is due to the presence of strontium sulfate in the fireworks' fuel.

Different types of chemical compounds when ignited create different colored fireworks.

Which metallic elements are likely present in the above fireworks display? What is the source of the energy that is absorbed by the metallic elements?

5.6
ATOMS GAIN OR LOSE ELECTRONS

Two major discoveries in the last hundred years have called for a revision of Boyle's definition of an element. The discoveries were that

• elements may gain or lose electric charges
• elements can emit high-energy particles and change spontaneously into other elements

Both of these discoveries conflict with Boyle's notion of an element.

An arrangement of 8 electrons in an energy level has special significance. Except for helium, all of the Group 18 elements (the noble gases) have 8 electrons in their outer energy level. These elements do not react easily and are considered to be very stable. The term stable octet is used for this stable configuration of 8 electrons in the outer energy level. Elements can either gain or lose electrons in order to have a stable octet.

A sodium atom with 11 electrons and 11 protons is said to be a neutral atom. The positive charges equal the negative charges. Electrons are placed outside the nucleus in their correct energy level.

In the first energy level, there are two electrons. In the second energy level there are 8. In the third there is 1 (see Figure 5.4). To become stable, an atom needs a complete outer energy level or 8 outer electrons. For sodium, this means giving away an electron to another atom.

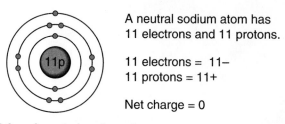

A neutral sodium atom has 11 electrons and 11 protons.

11 electrons = 11−
11 protons = 11+

Net charge = 0

Figure 5.4   A neutral sodium atom

The fluorine atom, which needs only 1 electron to complete its outer energy level, has 9 electrons: 2 in the first energy level and 7 in the second (see Figure 5.5). If it gains 1 more electron from an atom, such as sodium, it will have a complete second energy level.

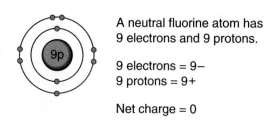

A neutral fluorine atom has 9 electrons and 9 protons.

9 electrons = 9−
9 protons = 9+

Net charge = 0

Figure 5.5   A neutral fluorine atom

When sodium gives away its outer electron, it becomes positively charged and is called a sodium ion. It has one less electron and a net charge of 1+ (see Figure 5.6).

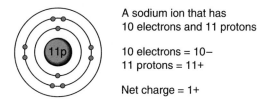

A sodium ion that has
10 electrons and 11 protons

10 electrons = 10–
11 protons = 11+

Net charge = 1+

Figure 5.6   A positively charged sodium ion

When fluorine accepts the electron that sodium has lost, it becomes negatively charged and it too, is called an ion. It has one extra electron and a net charge of 1- (see Figure 5.7). This ion is called a fluoride ion.

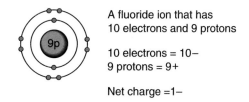

A fluoride ion that has
10 electrons and 9 protons

10 electrons = 10–
9 protons = 9+

Net charge =1–

Figure 5.7   A negatively charged fluoride ion

Try These

1. If an atom gives up an electron, what charge will it have?
2. If an atom accepts an electron, what charge will it have?

5.7
THE ATOMIC NUMBER

Atoms can be broken down by a high-speed accelerator into the particles that make them up. High-speed accelerators are nicknamed *atom smashers.*

When a nitrogen atom is smashed it breaks up into smaller fragments, as Figure 5.8 illustrates.

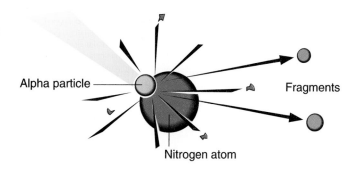

Alpha particle

Fragments

Nitrogen atom

Figure 5.8   A smashed nitrogen atom showing two fragments

The number of protons in the nucleus of an atom is called the atomic number and is given the symbol z. In a neutral atom, the number of protons is always equal to the number of electrons, or in other words, the charges are balanced. Every atom has its own z number. On the periodic table, each element's z number or atomic number is written in the top left corner of its box. For instance, hydrogen has a z number of 1, helium 2, and nitrogen 7.

Try These

1. What are the atomic numbers for the elements boron and carbon?
2. An atom has z=12. How many protons and electrons does it have?
3. The atomic number is always equal to the number of protons but not always to the number of electrons. Why?

5.8

THE ATOMIC MASS NUMBER

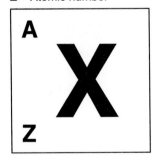

X = Element symbol
A = Atomic mass number
Z = Atomic number

The number of protons plus the number of neutrons in the nucleus of an atom make up its atomic mass number (A). The electron's mass is so small that it is not considered in the mass of the atom. Hydrogen, for example, has one proton and one electron but does not have any neutrons. Hydrogen is the only atom that has no neutrons. Its mass is one atomic mass unit or 1 u. The symbol for atomic mass is u. An atom of helium has two protons and two neutrons and a mass of 4 u.

In describing an element, the atomic number and atomic mass number are placed at the bottom left and top left of the symbol, respectively as shown in the margin. For example, the element fluorine has z=9 and A=19. Carbon has z=6 and A=12. Lithium has z=3 and A=7 (see Figure 5.9).

Figure 5.9   The atomic number and atomic mass number of the elements fluorine, carbon, and lithium

Since A is the number of protons and neutrons, and z is the number of protons, there must be 10 neutrons in any fluorine atom (A-z).

Atomic mass number (A) = Number of protons + number of neutrons
Atomic number (z) = Number of protons
Number of neutrons = Atomic mass number (A) – atomic number (z)

Try These

1. State the number of protons, electrons, and neutrons in a carbon atom and a lithium atom.
2. State the number of protons, neutrons, and electrons for the rest of the first 20 elements.
3. Prepare a table like the one shown below and fill in the blanks.

Element	$^{23}_{11}$Na				$^{40}_{18}$Ar
number of protons		13		9	
number of neutrons		14	16	10	
atomic mass number			31		
number of electrons					

4. If an atom of boron has 5 protons and 6 neutrons, what is its atomic mass in u?

SCIENCE & TECHNOLOGY

The Discovery of Element 109

It took six years to prepare for an experiment that took only 0.05 s to perform.

Element 109 was formed at 16:10, August 29, 1982 when the nucleus of one atom of iron sped down the atom smasher into a single atom of bismuth.

The nuclei of the two atoms fused, forming an atom of a new element. An instant later, the atom was gone. It was so unstable that it broke down into lighter elements. Due to its brief life, scientists could not determine how it behaved, but space has been reserved for it in today's periodic table of the elements.

High-speed accelerators can break down atoms.

5.9
ATOMIC THEORY THROUGH THE YEARS

The model of the atom has changed several times throughout history. With each important new discovery, scientists found it necessary to change their view of what atoms look like.

Our present atomic model is the result of years of work by many scientists. Table 5.1 summarizes the contributions of some important scientists toward an understanding of atomic structure.

Table 5.1   Summary of the development of the theory of atomic structure

Year	Scientist	Country	Contribution
500 B.C.	Democritus	Greece	first to suggest that matter consists of atoms
1809	Dalton	England	developed modern atomic theory
1886	Goldstein	Germany	discovered protons
1897	Thomson	England	discovered electrons
1898	Becquerel	France	discovered radioactivity
1910	Rutherford	England	discovered the nucleus
1913	Bohr	Denmark	suggested that electrons exist at various energy levels
1932	Chadwick	England	discovered neutrons

5.10
BOHR-RUTHERFORD DIAGRAMS

Bohr-Rutherford diagrams are drawings of atoms that show the number of protons and neutrons in the nucleus and the number of electrons in their energy levels.

Figure 5.10   The Bohr-Rutherford diagram of a hydrogen atom

The energy level nearest the nucleus, or the first energy level, can never hold more than two electrons at a time. If an atom has more than two electrons, a second energy level must be used. Lithium is such an element. It has three protons, three electrons, and four neutrons.

Figure 5.11   The Bohr-Rutherford diagram of a lithium atom

The second energy level can never hold more than eight electrons. Once the atom has more than ten electrons in total, a third energy level must be used. Sodium is such an element. It has 11 protons, 11 electrons, and 12 neutrons.

Figure 5.12   The Bohr-Rutherford diagram of a sodium atom

The third energy level can never hold more than 18 electrons. This would mean that once the atom had more than 28 electrons in total, a fourth energy level would be used. Complications in the model do occur. The energy levels are not always filled to the maximum before electrons begin to occupy the next energy level. This occurs first with potassium, which has 19 electrons. The first two energy levels are filled but there are only eight electrons in the third energy level and one electron in the fourth energy level. An arrangement with eight electrons in an energy level seems to have special significance.

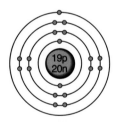

Figure 5.13   The Bohr-Rutherford diagram of a potassium atom

Table 5.2   Maximum number of electrons per energy level

Level	Maximum Number of Electrons
1	2
2	8
3	18

Try These

1. Draw Bohr-Rutherford diagrams for the following atoms:
   a. helium
   b. beryllium
   c. oxygen
   d. aluminum
   e. argon

5.11
ISOTOPES

The simplest atom, the hydrogen atom, has only two of the three basic subatomic particles. It has one proton and one electron. It is the only atom that does not have a neutron.

However, hydrogen has two closely related relatives. They occur if either one or two neutrons are added to the nucleus of a hydrogen atom. This addition of one neutron or two neutrons results in an additional two types of hydrogen atoms, each with different masses.

If one neutron is added to hydrogen, a heavier nucleus results. A new atom called deuterium is formed, having a mass of 2 u, as illustrated below.

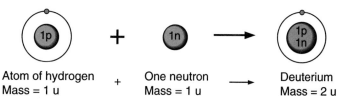

Atom of hydrogen   +   One neutron   Deuterium
Mass = 1 u   Mass = 1 u   Mass = 2 u

Figure 5.14   A deuterium atom

Hydrogen has an even rarer relative, tritium. Tritium is formed by the addition of two neutrons to the nucleus of a hydrogen atom, as the following illustrates.

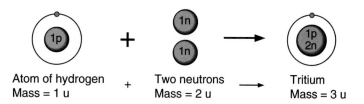

Atom of hydrogen   +   Two neutrons   Tritium
Mass = 1 u   Mass = 2 u   Mass = 3 u

Figure 5.15   A tritium atom

In each type of hydrogen there is one proton and one electron. The only difference for each of hydrogen's relatives is the mass caused by the addition of one or more neutrons. A special name, isotopes, is given to atoms of the same element that have different masses. Some isotopes are unstable or radioactive atoms. This means that they spontaneously change into other isotopes. Radioactive isotopes, or radioisotopes, have been used extensively in biological research and in medical treatment of certain diseases.

Try These

1. a. Helium has two naturally occurring isotopes. They are ^3_2He and ^4_2He. How many protons, electrons and neutrons does each atom have?

   b. What is the difference between the two atoms?

2. a. Neon has three isotopes that occur naturally. They are $^{20}_{10}\text{Ne}$, $^{21}_{10}\text{Ne}$, and $^{22}_{10}\text{Ne}$. Draw Bohr-Rutherford diagrams for each.

   b. How many protons, electrons, and neutrons does each different atom have?

   c. What is the difference among the three isotopes?

5.12 ACTIVITY

MASS SPECTROMETRY

Chemists have used different methods to determine the mass of each of the elements. The most recent method is called mass spectrometry. If two balls of different masses, one metal and the other polystyrene foam, were dropped from the same height with no wind present, they would land at a point directly below the point from which they were dropped. If, however, there was a wind blowing, the polystyrene foam ball would be blown farther off course than the metal ball. In a mass spectrometer, the magnetic field functions as the wind did with the ball. The difference in how atoms move through an electric field is due to the mass of each atom.

In a mass spectrometer, the atoms are given a positive charge by removing an electron. They then begin to move toward a negatively charged electrode and pass through a magnetic field. The magnetic field acts as a deflecting source. The lighter atoms are deflected more than the heavier ones.

In this activity you will construct a model of a mass spectrometer.

Materials

glass or clear plastic plate (1.0 m x 1.5 m)
leveling supports
magnet
launcher (inclined plane)
collector boxes
ball bearings of various sizes

Method

1. Set up your materials as shown in Figure 5.16. Use the leveling supports to make sure that the apparatus is level.
2. Roll the ball bearings down the launcher one at a time.
3. Allow the ball bearings to collect in the collector boxes.
4. Observe which ball bearings collected in each box.

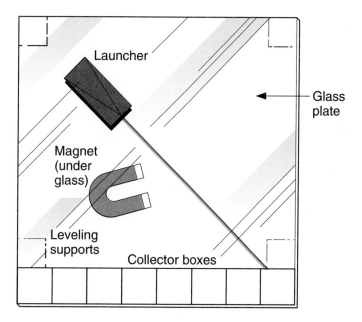

Figure 5.16   A model of a mass spectrometer

Questions and Conclusions

1. What relationship exists between the mass of each ball bearing and the amount it was deflected?
2. Imagine that the ball bearings are atoms. Should atoms of the same element always land in the same collector box?
3. Would isotopes of the same element always land in the same collector box or in different collector boxes?
4. If you were the scientist setting up this experiment, how would you relate each collector box to different atomic masses?

CAREER OPPORTUNITY

X-ray Technician

X-ray technicians are an essential part of the health-care team working in hospitals and medical clinics. Their work is very interesting and challenging.

Radiology technicians, as they are sometimes called, have completed a two-year college training program that includes courses in anatomy, physiology, and radiographic techniques. Their job is to prepare patients for X rays and to operate X-ray equipment. Much of their work is done under emergency conditions in the emergency unit of a hospital.

X-ray technicians also assist in treating specific diseases by exposing patients to concentrated doses of X rays. The technician must be aware of the potential hazards of radiation and know how to operate the equipment safely.

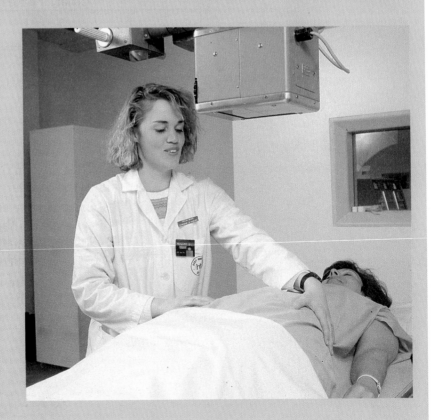

Exercises

Key Words

Review

1. How did Niels Bohr change the model of the atom?
2. Draw Bohr-Rutherford diagrams for
   a. oxygen   b. sulfur   c. potassium   d. chlorine
3. Identify the neutral atoms in the following Bohr-Rutherford diagrams:

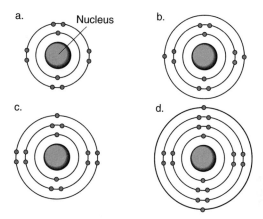

4. What two discoveries led to the revision of Boyle's definition of an element?
5. List the three subatomic particles present in an atom.
6. Define and give an example of an isotope.
7. How does a deuterium atom differ from a hydrogen atom?
8. Copy the chart shown below and fill in the missing information.

Nucleus	$^{40}_{20}$Ca	$^{56}_{26}$Fe			$^{238}_{92}$U
number of protons				99	
number of neutrons			31	153	
atomic mass number			57		
number of electrons					

9. What instrument is used to measure the relative masses of atoms?

Extension

1. Research the story of Niels Bohr's life. Why was his work so important during World War II?

CHAPTER 6
ELEMENTS AND
PERIODICITY

Elements make up all matter in our bodies and in everything around us. Elements are contained in products that we use every day and many elements are essential to the life processes of all living things.

In this chapter you will trace the discovery and naming of the elements. You will discover more about the subatomic particles that make up the smallest component of an element—the atom.

You will also learn how elements are classified into groups or families according to their behavior. These groups are then arranged to form a table called the periodic table.

This chapter also discusses the properties of metals and nonmetals, and compares the reactivity of elements within the same group.

Key Ideas

- Chemistry has a universal language of symbols and nomenclature.
- Elements each have their own properties. No two elements are the same.
- Every atom has a particular atomic mass and atomic number.
- Elements are arranged in increasing atomic number in the periodic table.
- Elements in the same group have similar properties.
- Properties of an element can be inferred by knowing the trends within a group in the periodic table.

6.1

SYMBOLS AND NAMES

☾̇	Antimony
Β Ш ᶜ 8	Bismuth
♀	Copper
♂	Iron
☉	Gold
♄	Lead
☿	Mercury
☽ ♁	Silver
♃	Tin
Z₍	Zinc

Figure 6.1   Early symbols for some elements

Jöns Berzelius (1779–1848)

Chemical symbols are used to convey ideas quickly and concisely. These shorthand notations are merely a convenience and contain no mysterious concepts that cannot be expressed in words.

It is interesting to read how the elements acquired their names. Believe it or not, scientists have never named an element they discovered after themselves. The names given to elements can

• honor other scientists. Curium was named after Marie Curie, the scientist responsible for discovering radioactivity.

• indicate where the element was discovered. Americium was discovered in America.

• describe the element's properties. Krypton is Greek for *"hidden one."* The element was given this name because it is very rare.

As more and more elements were discovered, the need became evident for an organized system for discussing these elements among scientists in different countries. Scientists all over the world began to record their experimental results in journals so that these results could be read by all other interested scientists. It was important that all of the journals used common symbols to represent the elements. Some of the early symbols for the elements are shown in Figure 6.1.

A Swedish chemist by the name of Jöns Berzelius recognized the need for a simpler system to represent all the elements. In 1814, he devised a system that is now used by scientists all over the world.

Berzelius took the first letter of each element and capitalized it: **H** represents hydrogen, **U** represents uranium, and **C** represents carbon. However, Berzelius encountered two problems. First, there were over a hundred elements to name and only 26 letters in the alphabet. Secondly, the names of some elements began with the same letter.

This problem was solved for elements beginning with the same letter by capitalizing the first letter and following it with the second lower-case letter. As a result, most elements now have two-letter symbols. Some examples are **He,** helium, **Ba,** barium, and **Be,** Beryllium.

Some elements have symbols other than the first letters of their name. The clue to most of these symbols is found in Latin, a language often used for scientific names.

Table 6.1   Symbols with Latin derivatives

Element	Symbol	Latin Name
antimony	Sb	stibium
copper	Cu	cuprum
gold	Au	aurum
iron	Fe	ferrum
lead	Pb	plumbum
mercury	Hg	hydrargyrum
potassium	K	kalium
silver	Ag	argentum
sodium	Na	natrium
tin	Sn	stannum

Try These

1. Study the list of symbols in Figure 6.1. On a separate sheet, list all the names and symbols without referring to the figure. Did you have any difficulty doing it?
2. How many elements begin with the letter C?
3. Name two elements that are named after the country where they were discovered.
4. Name one element that is named after a famous scientist.

6.2 ACTIVITY
PICK AN
ELEMENT –
ANY ELEMENT

Method

1. Choose an element from the periodic table and use the library resource center to find the following:

   Discovery
   a. What year was the element discovered?
   b. What is the name of the scientist who discovered the element?
   c. What is the story of its discovery?

   Sources
   d. What chemical compounds are the source of this element in nature?
   e. Where are these compounds found?
   f. How are the source compounds collected?
   g. How is the element removed from the source compounds it is found in?

Uses

h. List several uses of the element. For example, is it used in industry, in agriculture, or in the household?

i. In what form is the element used? Is it used as the pure element or in a compound?

2. Build a three-dimensional model of an atom of the element you chose. Display it in the classroom.

3. Present your display and research as a report to your teacher and/or as a presentation to your class.

6.3
ELEMENTS
SURROUND US

In addition to making up all living things, elements exist all around us in the atmosphere and the oceans. Most elements exist in molecular or ionic form. The elements most readily available are those occurring in the atmosphere.

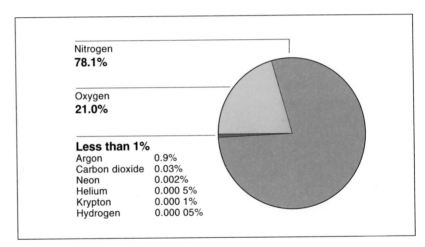

Nitrogen
78.1%

Oxygen
21.0%

Less than 1%
Argon	0.9%
Carbon dioxide	0.03%
Neon	0.002%
Helium	0.000 5%
Krypton	0.000 1%
Hydrogen	0.000 05%

Figure 6.2   Composition of clean, dry air

These gases can be isolated by liquefying pure air. Once a liquid forms, each gas can be extracted or removed by distilling the liquid air. All of the neon used in neon signs is obtained this way, along with large amounts of pure oxygen.

The next most widespread source of the elements is the ocean. Most elements in seawater are found as electrically charged particles or ions. The most common are sodium and chloride ions, which give seawater its characteristic saltiness. In addition to those listed on page 101, almost every element found on land is also present in the oceans. As well, various amounts of the gases that occur in the atmosphere are found dissolved in the oceans.

Table 6.2   Dissolved substances in seawater

Ion	g/kg Seawater
chloride (Cl^-)	18.980
sodium (Na^+)	10.556
sulfate (SO_4^{2-})	2.649
magnesium (Mg^{2+})	1.272
calcium (Ca^{2+})	0.400
potassium (K^+)	0.380
bicarbonate (HCO_3^-)	0.140
bromide (Br^-)	0.065
boric acid (H_3BO_3)	0.026
strontium (Sr^{2+})	0.013
fluoride (F^-)	0.001

As you become more comfortable with the language and symbols of chemistry, you will become more aware of the composition of substances that exist everywhere around you.

Table 6.3 lists the principal elements that compose the human body and their biological roles.

Table 6.3   Principal elements of the human body

Element	Percentage by Mass	Biological Role
oxygen	62.00	component of water, carbohydrates
carbon	20.00	backbone of all organic molecules
hydrogen	10.00	component of water, other molecules
nitrogen	3.00	component of proteins
calcium	2.50	major component of bones and teeth
phosphorus	1.14	required for biological energy
chlorine	0.16	component of stomach acid
sulfur	0.14	component of proteins
potassium	0.11	required for nerve and muscle action
sodium	0.10	required for nerve and muscle action
magnesium	0.07	required for nerve and muscle action
iodine	0.01	component of thyroid hormone
iron	0.01	component of blood

Try These

1. A wide variety of common products contain chemical compounds. To see what elements are present in them, check the labels on products in your kitchen, bathroom, and laundry. Make a table in your notebook similar to the one below and list the substance, the chemicals it contains, and determine as many of the elements in the product as you can. One example is done for you.

Common Substance	Chemical Name	Elements
bleach	sodium hypochlorite	sodium, chlorine, oxygen

2. Name as many elements as you can that are present in the common household products represented below.

a. **SALT**

FREE
RUNNING
TABLE
SALT

INGREDIENTS; SALT, SODIUM SILICO ALUMINATE, POTASSIUM IODIDE.

b. Magnesia

TABLETS

Each Tablet contains:
310 mg magnesium hydroxide.

c. "This product contains sodium fluoride which is, in our opinion, an effective decay preventive agent, and is of significant value when used in a conscientiously applied program of oral hygiene and regular professional care."

Sodium fluoride 0.243%

d. PAIN RELIEVER & ANTACID

Adult Dose: Two tablets dissolved in water every 4 hours as required up to a maximum of 8 tablets in 24 hours.

Do not exceed this recommended maximum dose or continue use of this product for more than 5 days unless advised by a physician.

Consult a physician before taking this product during the last 3 months of pregnancy, when nursing, or if you are on a sodium (salt) restricted diet.

A physician should be consulted before giving this medicine to children or teenagers with chicken pox or influenza.

Keep safely out of reach of children.

Each tablet contains: Acetylsalicylic acid 325 mg, heat treated sodium bicarbonate 1916 mg and citric acid 1000 mg.

Active ingredients in water include ions of: acetylsalicylate 323 mg, sodium 554 mg and citrates 985 mg.

6.4

THE EARLY PERIODIC TABLE

The first periodic table of the elements was proposed by a Russian chemist, Dmitri Mendeleev, in 1869. He arranged the elements in order of increasing atomic mass. When he did this, he noticed some periodic (recurring) properties. For example, lithium, sodium, and potassium all react violently when placed in water. He grouped all the elements with similar properties into vertical groups called families. From this, Mendeleev formulated a periodic law, which stated that the properties of the elements are a periodic function of their atomic mass. As soon as isotopes were discovered, this law had to be changed.

Row	Group I	Group II	Group III	Group IV	Group V	Group VI	Group VII	Group 0
1	H = 1							
2	Li = 7	Be = 9	B = 11	C = 12	N = 14	O = 16	F = 19	
3	Na = 23	Mg = 24	Al = 27	Si = 28	P = 31	S = 32	Cl = 35.5	
4	K = 39	Ca = 40	Sc = 44	Ti = 48	V = 51	Cr = 52	Mn = 55	Fe = 56 Co = 58.5 Ni = 59
5	Cu = 63	Zn = 65	Ga = 70	Ge = 72	As = 75	Se = 79	Br = 80	
6	Rb = 85	Sr = 87	Y = 89	Zr = 90	Nb = 94	Mo = 96		Ru = 103 Rh = 104 Pd = 106
7	Ag = 108	Cd = 112	In = 113	Sn = 118	Sb = 120	Te = 125	I = 127	
8 9	Cs = 113	Ba = 137	La = 138	Ce = 140				
10			Yb = 173		Ta = 182	W = 184		Os = 191 Ir = 193 Pt = 196
11	Au = 198	Hg = 200	Tl = 204	Pb = 206	Bi = 208			
12				Th = 232		U = 240		

Figure 6.3   Mendeleev's Table

Try These

1. State the periodic law.
2. Which of the elements do not follow Mendeleev's table?

6.5

THE MODERN PERIODIC TABLE

The modern periodic table is not arranged in order of increasing atomic mass, but in order of increasing atomic number, or the number of protons in the nucleus (see Figure 6.4 on pages 104–105). The first major division in the table is between the metals and nonmetals. All elements to the left of the line that borders **B, Si, As, Te,** and **At** are metals. This dividing line is often referred to as the staircase. All elements to the right of the staircase are nonmetals.

Continued on page 106

Periodic Table of the Elements

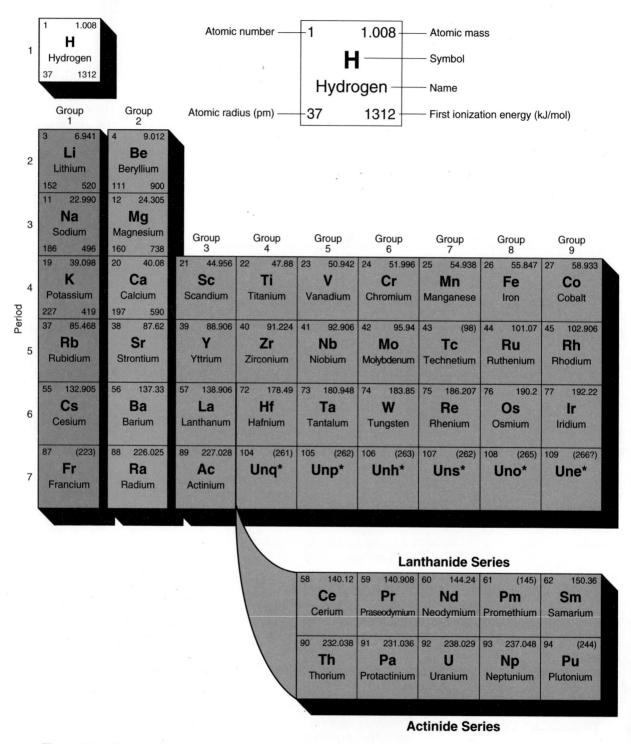

Figure 6.4   Periodic Table of the Elements. Mass numbers in parentheses are those of the most stable or most common isotope.

*The systematic names and symbols for elements of atomic number greater than 103 will be used until the approval of trivial names by IUPAC.

Metals
- Alkali metals
- Alkaline-earth metals
- Transition metals
- Other metals

Metalloids
- Metalloids

Nonmetals
- Halogens
- Other nonmetals

Noble Gases
- Noble gases

Group 10	Group 11	Group 12	Group 13	Group 14	Group 15	Group 16	Group 17	Group 18
								2 4.003 **He** Helium 50 2372
			5 10.81 **B** Boron 79 801	6 12.011 **C** Carbon 77 1086	7 14.007 **N** Nitrogen 74 1402	8 15.999 **O** Oxygen 73 1314	9 18.998 **F** Fluorine 71 1681	10 20.179 **Ne** Neon 65 2081
			13 26.982 **Al** Aluminum 143 578	14 28.086 **Si** Silicon 117 787	15 30.974 **P** Phosphorus 116 1012	16 32.06 **S** Sulfur 102 1000	17 35.453 **Cl** Chlorine 99 1251	18 39.948 **Ar** Argon 95 1521
28 58.69 **Ni** Nickel	29 63.546 **Cu** Copper	30 65.39 **Zn** Zinc	31 69.72 **Ga** Gallium	32 72.59 **Ge** Germanium	33 74.922 **As** Arsenic	34 78.96 **Se** Selenium	35 79.904 **Br** Bromine	36 83.80 **Kr** Krypton
46 106.42 **Pd** Palladium	47 107.868 **Ag** Silver	48 112.41 **Cd** Cadmium	49 114.82 **In** Indium	50 118.71 **Sn** Tin	51 121.75 **Sb** Antimony	52 127.60 **Te** Tellurium	53 126.905 **I** Iodine	54 131.29 **Xe** Xenon
78 195.08 **Pt** Platinum	79 196.967 **Au** Gold	80 200.59 **Hg** Mercury	81 204.383 **Tl** Thallium	82 207.2 **Pb** Lead	83 208.980 **Bi** Bismuth	84 (209) **Po** Polonium	85 (210) **At** Astatine	86 (222) **Rn** Radon

63 151.96 **Eu** Europium	64 157.25 **Gd** Gadolinium	65 158.925 **Tb** Terbium	66 162.50 **Dy** Dysprosium	67 164.930 **Ho** Holmium	68 167.26 **Er** Erbium	69 168.934 **Tm** Thulium	70 173.04 **Yb** Ytterbium	71 174.967 **Lu** Lutetium
95 (243) **Am** Americium	96 (247) **Cm** Curium	97 (247) **Bk** Berkelium	98 (251) **Cf** Californium	99 (252) **Es** Einsteinium	100 (257) **Fm** Fermium	101 (258) **Md** Mendelevium	102 (259) **No** Nobelium	103 (260) **Lr** Lawrencium

Along the top of the periodic table are the numbers 1 to 18. These vertical columns in the periodic table are called groups or families. Each member of the group behaves in somewhat the same way chemically.

Along the left side of the periodic table are the numbers 1 to 7. These divisions are periods and the elements are arranged in increasing atomic number from left to right.

The periodic table is set up as a cycle. As you go across the table from left to right, the behavior of the elements changes. But as you return to the left side again, the pattern of behavior repeats.

There are other divisions in the periodic table, which are listed in Figure 6.4.

6.6 ACTIVITY
BOHR-RUTHERFORD DIAGRAMS AND THE PERIODIC TABLE

In Chapter 5 you learned how to draw Bohr-Rutherford diagrams. In this activity you will use Bohr-Rutherford diagrams to see some of the recurring or periodic properties of the elements.

Materials

periodic table
20 square sheets of cardboard (5 cm x 5 cm)

Front

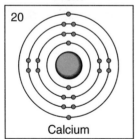

Calcium

Back

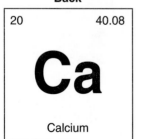

Method

1. Number the front of each of your 20 squares from 1 to 20, placing the number in the top left-hand corner as shown in the margin.
2. At the bottom of the square, write the name of the element that corresponds to that number.
3. In the center, draw a Bohr-Rutherford diagram for each element.
4. Turn each card over and print the name, symbol, mass, and atomic number for each element on the back.

Questions and Conclusions

1. Consider the diagrams for the elements **H, Li, Na,** and **K.** What do they have in common?
2. Consider the diagrams for the elements **Be, Mg,** and **Ca.** What do they have in common?

3. In general, what can be said about the following groups of elements?

   B, Al   C, Si   N, P   O, S   F, Cl   He, Ne, Ar

4. Arrange your squares into groups according to the one property they have in common. At the top of each group, label 1 for Group 1, 2 for Group 2, and so on.

5. Consider the diagrams for the elements **H** and **He**. What do they have in common?

6. Consider the diagrams for the elements **Li, Be, B, C, N, O, F, Ne**. What do they have in common?

7. In general, what can you conclude about the elements that are arranged horizontally?

8. Turn each square over onto its back. How are the elements arranged from left to right?

9. Elements react according to the number of electrons in the outer shell. What is meant by the word "periodic" in periodic table?

6.7 ACTIVITY
REACTIVITY
AND THE
PERIODIC TABLE

Elements in the periodic table are arranged in groups. Each group has its own chemical behavior. In this activity you will be comparing elements within a group as well as comparing groups of elements.

The behavior of an element can often be shown by reacting it with another substance. You will be reacting Group 1 and Group 2 elements with water and observing their reactions. You will be reacting Group 17 elements with other Group 17 ions from some compounds. The single element is considered more reactive if it can displace, or kick out, the element in the compound. Since chlorine can displace bromine, it is considered more reactive.

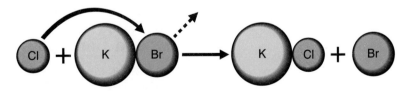

chlorine + potassium bromide ⟶ potassium chloride + bromine

Caution: Bromine water and chlorine water are corrosive and cause burns. Handle them with care. Wear goggles and an apron. If you spill any on yourself, rinse it off immediately with large amounts of cold water and have your partner tell your teacher.

Perform steps 3 and 4 in a fume hood.

Materials

test tubes (9)	potassium bromide
test tube rack	potassium iodide
250 mL beakers (2)	magnesium
watch glass	calcium
potassium`	bromine water
sodium	chlorine water
lithium	iodine water
potassium chloride	disposable vinyl gloves

Method

1. Your teacher will demonstrate the first three reactions. Observe what happens when small pieces of potassium, lithium, and sodium are dropped into 100 mL of tap water.
2. Obtain a small piece of magnesium and a small piece of calcium from your teacher. Add each, in turn, to 100 mL of tap water in a 250 mL beaker. Cover with a watch glass and observe. Use fresh water each time.
3. a. Number three test tubes 1, 2, and 3.
   b. Place a few crystals of potassium chloride in test tube 1, potassium bromide in test tube 2, and potassium iodide in test tube 3.
   c. Add 3 mL of tap water to each test tube and shake until the crystals have dissolved completely.
   d. Add 1 mL of chlorine water to each test tube and observe any color changes.
4. Repeat step 3 using bromine water.
5. Repeat step 3 using iodine water.
6. Prepare a data table summarizing your results.

Questions and Conclusions

1. Arrange **Li, Na,** and **K** in order of the least reactive to the most reactive.
2. If this list could be extended to include all of the other elements in Group 1, what would be the order from most reactive to least reactive?
3. Compare your results for **K** and **Ca,** and **Na** and **Mg.** Which group is more reactive?
4. Which metal is the most reactive metal listed in the periodic table?

5. List the three nonmetals **Cl, Br,** and **I** in order of least reactive to most reactive.

6. Which do you think is the most reactive nonmetal in the periodic table?

6.8 ACTIVITY
GRAPHING
PERIODIC TRENDS

Atoms become ions by gaining or losing one or more electrons. Ionization energy is the energy required to remove the outermost electron from a neutral atom. In this activity you will find out how atomic radius and ionization energy are related to each other by drawing graphs for atomic radius and ionization energy against the atomic number of each of the first 20 elements. Atomic radius is measured in picometers (1 pm = 10^{-12} m). Ionization energy is measured in kJ/mol.

Graphing is an important scientific skill. You should take care that your graphs are neatly and accurately constructed. You can then examine them to see if there are regular patterns or variations in the properties that you are investigating.

Materials

graph paper   pencil

ruler   periodic table

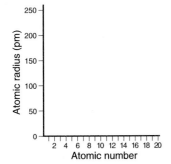

Method

1. The horizontal axis must have twenty equally-spaced marks, one for each of the first twenty elements. Label this axis Atomic number.

2. Divide the vertical axis into intervals so that the largest number of the property you are graphing is near the top of the page. Label the vertical axis with the name of the property, for example, Atomic radius.

3. Plot the atomic radius of each of the first 20 elements against its atomic number on one sheet of graph paper.

4. Repeat step 3 for ionization energy against atomic number, using a separate sheet of graph paper.

5. Connect the points to complete the graph.

Questions and Conclusions

1. a. What is the name of a horizontal row of the periodic table?
   b. What is the name of a vertical column of the periodic table?

2. a. Compare the section of the graph for elements 3 through 10 with the section for elements 11 through 18. How are these sections similar? How are they different?
   b. What is the general trend in atomic radii within a period?

3. a. Label the points on the graph with the symbols of the elements as shown below.

Atomic Number	Symbol
2	He
3	Li
10	Ne
11	Na
18	Ar
19	K

   b. What is the trend in atomic radii within a group as you go from **Li** to **Na** to **K**?
   c. What is the trend in atomic radii as you go within a group from **He** to **Ne** to **Ar**?
   d. What is the trend in atomic radii within a group?

4. What is the trend in ionization energy as you go across a period?

5. What is the trend in ionization energy within a group?

6. Which group of elements loses its electrons most easily?

7. Which group of electrons holds on to its electrons most tightly?

8. Draw Bohr-Rutherford diagrams for helium, neon, and argon. What do you notice about the energy level of each? Why do these elements have such high ionization energies?

6.9 ACTIVITY
PROPERTIES
OF METALS
AND NONMETALS

There are many everyday uses of metals that are familiar to you. However, you may not be as aware of some common uses of nonmetals.

Table 6.4   Common uses of nonmetals

Element	Use
neon	neon lights
iodine	medicine
sulfur	fertilizer
phosphorus	matches, fertilizer
helium	balloons
chlorine	water treatment

Materials

test tubes (6)   magnesium pellets/strips
test tube rack   powdered sulfur/lump of sulfur
conductivity apparatus   charcoal/graphite rod
copper pellets/strips   steel wool (iron)/iron rod
zinc pellets/strips   hydrochloric acid

Caution: Hydrochloric acid is corrosive and causes burns. Handle it with care. Wear goggles and an apron. If you spill any on yourself, rinse it off immediately with large amounts of cold water and have your partner tell your teacher.

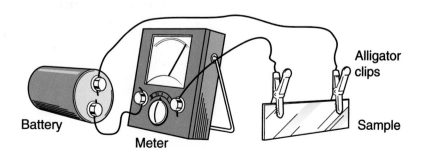

Figure 6.5   Apparatus to test for conductivity

Method

1. Study the properties of metals and nonmetals listed in Table 6.5.
2. Record the color of each element.
3. Is each element malleable? Record your results.
4. Check each element for luster. Record your results.
5. Place a small piece of each element into separate test tubes. Add hydrochloric acid to each to a depth of 2 cm. Check for formation of bubbles.
6. Prepare a chart summarizing your results.
7. Test each sample with a conductivity apparatus.

Table 6.5   A summary of the properties of metals and nonmetals

Metals	Nonmetals
good conductors of heat	poor conductors of heat
good conductors of electricity	poor conductors of electricity
lustrous – reflect light	dull – do not reflect light
malleable – can be hammered into sheets	brittle – cannot be hammered into sheets
ductile – can be drawn into wires	not ductile – cannot be drawn into wires

Questions and Conclusions

1. From your observations, which of the elements are metals and which are nonmetals?
2. Define the terms "lustrous" and "malleable."
3. a. List three characteristic properties of a metal.
   b. List three characteristic properties of a nonmetal.
4. a. Which side of the periodic table are the metals on?
   b. Which side of the periodic table are the nonmetals on?
5. What is another word meaning
   a. shiny?
   b. a substance listed in the periodic table?
   c. can be bent?
6. What do metals conduct very well?

There are a few exceptions to the rule that nonmetals are nonconductors. These elements are called metalloids (metal-like), but most of their properties resemble those of nonmetals. Silicon, for example, has some metallic properties and non-metallic properties. Why is it referred to as a semiconductor? Carbon is a nonmetal, but in another form (graphite), it has a single metallic property. From your observations in Activity 6.9, what metallic property does carbon have?

SCIENCE & TECHNOLOGY

Silicon—An Element for the Electronic Age

Tiny silicon "chips" such as this one are used in computers and other electronic equipment.

Silicon is found as various silicates and as silicon dioxide (SiO_2) in sand, quartz, clay, and igneous rock. These compounds make up 87% of the earth's crust. Silicon is the second most abundant element in the earth's crust.

The properties of silicon compounds have been used for centuries. In its crystalline structure, silica or SiO_2 (silicon dioxide), has been put to good use making such diverse products as glass, mortar, and concrete.

In their hexagonal plate structure, silicates act like layers of paper lying on top of one another that can slide easily over one another. This property can be observed in mica and talcum powder, which are examples of this form. Clay is another example of this plate structure. Water molecules between the layers make the clay slippery. As clay dries or as it is "fired," it becomes very hard and strong as bonds are formed between the layers. Fired clays are called ceramics.

Chains of silicon and oxygen can also be formed, producing silicones. Short chain silicones are liquids and longer chain silicones are rubber-like solids. These are useful because they repel water and resist chemicals.

Silicon has become widely known for a much different reason. Silicon is a metalloid element. It has some of the properties of metals and some of the properties of nonmetals. It has a metallic luster but it is an insulator. Silicon is useful in electronics because it can be made into a semiconductor. This is accomplished by adding small amounts of impurities to pure silicon. This process, called doping, is used to increase the conductivity of silicon.

6.10
THE ALKALINE EARTH METALS

The elements beryllium, magnesium, calcium, strontium, barium, and radium make up the group of elements called alkaline earth metals. The two most common elements of this group are calcium and magnesium. Table 6.6 lists their forms and uses.

Table 6.6   The uses of calcium and magnesium

Element	Forms Found	Applications
Mg	metallic	light alloys
	sulfate	fire proofing
	chloride	floor-sweeping compounds
	silicate	absorbent for oil spills
Ca	sulfate	plaster of paris
	phosphate	fertilizer
	oxide	cement, mortar
	other salts	healthy bones, teeth

All of the members of the alkaline earth metals are very reactive. When calcium is added to water it reacts vigorously to give off hydrogen gas.

calcium + water ⟶ calcium hydroxide + hydrogen

The flame tests you performed in Chapter 5 showed that magnesium burns with a brilliant white light and calcium burns with a red flame. The word equations for these reactions are

magnesium + oxygen ⟶ magnesium oxide
calcium + oxygen ⟶ calcium oxide

Both magnesium and calcium react with acid as follows:

magnesium + hydrochloric acid ⟶ magnesium chloride + hydrogen gas
calcium + hydrochloric acid ⟶ calcium chloride + hydrogen gas

Calcium and magnesium behave in the same way. Both are in the same group.

When magnesium and calcium react with air, oxides are formed. If these oxides are added to water, a base is formed.

magnesium oxide + water ⟶ magnesium hydroxide (a base)

calcium oxide + water ⟶ calcium hydroxide (a stronger base than magnesium hydroxide)

Magnesium hydroxide is an ingredient in some patent liquid medicines that are used to neutralize acids in the stomach. Calcium hydroxide is also a base and is an ingredient in limewater. Limewater has many uses, but it is mainly used to neutralize undesired acids in soil.

6.11 ACTIVITY COMPARING THE REACTIVITY OF CALCIUM, MAGNESIUM, AND THEIR SALTS

In this activity you will examine the reactivity of two alkaline earth metals, calcium and magnesium.

Materials

burner or hot plate	calcium carbonate
tongs	magnesium ribbon
test tubes (6)	calcium oxide
test tube rack	magnesium carbonate
Erlenmeyer flask	dilute hydrochloric acid
250 mL beaker	phenolphthalein indicator
glass stirring rod	calcium sulfate
medicine dropper	limewater
calcium metal	penny

Caution: Do not look directly at the light from burning magnesium. View it through cobalt blue glass.

Calcium is corrosive and causes burns. Handle it with care. Wear goggles and an apron. If you spill any on yourself, rinse it off immediately with large amounts of cold water and have your partner tell your teacher.

Method

1. Drop a small piece of calcium metal into a beaker half full of tap water. Record what happens.
2. Half fill an Erlenmeyer flask with water and bring it to a boil. When steam is being produced, gently lower a lit strip of magnesium ribbon into the steam, using tongs. Be careful not to let the burning magnesium touch the beaker. Does the magnesium still burn?
3. Light a 4 cm strip of magnesium and let it burn over the mouth of a test tube. When it has finished burning, place the residue in 5 mL of water in a test tube. Stir the mixture. Is the product soluble in water? Add a drop of phenolphthalein. What do you observe?
4. Add a 1 g sample of calcium oxide to a test tube with 10 mL of water and three drops of phenolphthalein. What do you observe? Is the substance soluble in water?
5. Place magnesium carbonate to a depth of 2 cm in a test tube and heat strongly. Hold a stirring rod with a drop of limewater in the test tube. What happens to the lime-water? What gas is given off?
6. Repeat step 5 using calcium carbonate in place of magnesium carbonate.
7. Place 1 cm^3 of calcium carbonate in a test tube. Add three drops of dilute hydrochloric acid. What do you observe? Hold a stirring rod with limewater on the tip in the test tube. What happens to the limewater? What gas is given off?
8. Take a little calcium sulfate and heat it strongly in a test tube. What gas is given off? Describe the solid left in the test tube. Add a little water to the powder left in the test tube and stir until a paste forms. Remove the paste. Press a penny into the paste and allow to stand for 30 min. Remove the penny. Examine the impression. Suggest some uses for this compound.

Magnesium burning

Questions and Conclusions

1. What properties do magnesium and calcium have in common?
2. With reference to their properties, explain the location of magnesium and calcium in the periodic table.
3. Predict how strontium would react with
   a. water   b. acid
4. Summarize the properties of the elements in Group 2.

6.12
HARDNESS
OF WATER

You have studied the reactivity of magnesium and calcium and also looked at some practical applications. One place where the alkaline earth metals prove to be a menace is in water supplies, requiring the use of much more soap when washing. If there is a large deposit of calcium chloride, calcium sulfate, or any of the other salts mentioned in Table 6.6 in the bedrock, these salts eventually get into our water supply. When this happens, the water is called hard water.

Ions such as calcium, magnesium, and iron can cause water to have undesirable characteristics. If water contains more than two hundred parts per million of these ions, it is considered hard. Hard water is not very good for washing dishes or clothes or for bathing. A bathtub ring is formed when soap is used in hard water. The residue or scum forms when calcium reacts with one of the ingredients in the soap.

There are two types of water hardness. Hardness caused by hydrogen carbonates is called temporary hardness and can be removed by boiling the water. Hardness caused by sulfate and chloride salts is called permanent hardness because it cannot be removed by boiling.

Water softeners eliminate hard water. They contain minerals called zeolites, which act as ion exchangers. As the hard water passes over the zeolites, the unwanted calcium, magnesium, and iron ions are exchanged for sodium ions. The sodium ions come from a large supply of sodium chloride (salt) which is stored in the water softener tank. The exchange happens because the calcium ions form an ionic compound with the zeolites.

Much of this problem has been eliminated by the development of synthetic detergents. They provide excellent washing conditions, even in hard water.

Try These

1. What would you predict the hardness of water to be in your area?
2. Why does scale sometimes build up in kettles? What can be done to remove it?

6.13 ACTIVITY
TESTING
THE HARDNESS
OF WATER

In this activity you will be determining the hardness of the water in your area. You will do this by determining the amount of soap required to make a lather. In areas where the water is soft, very little soap is required to form a rich lather. In areas where water is hard, more soap is required.

Materials

burner or hot plate	burette
Erlenmeyer flask	pipette
stopper	medicine dropper
250 mL beaker	distilled water
watch glass	soap solution

Method

1. Boil 100 mL of tap water for 15 min. Allow it to cool.
2. Place ten drops of tap water on a clean watch glass placed on a beaker of boiling water. When the water in the watch glass has evaporated, describe what is left.
3. Fill a burette with soap solution. Use a pipette to transfer 25 mL of tap water into the Erlenmeyer flask. Add a soap solution to the flask 1 mL at a time. After each addition, stopper the flask and shake. When a lather remains for 2 min after shaking, the process (called a titration) is complete. Record the volume of soap used.
4. Repeat step 3, replacing the tap water with distilled water.
5. Use a pipette to transfer 25 mL of the water from step 1 into the Erlenmeyer flask and repeat step 3.
6. Prepare and complete a chart similar to the one shown.

Solution	Soap Added (mL)	Averages
a. tap water		
b. distilled water		
c. boiled water		
permanent hardness	a. minus b.	
temporary hardness	a. minus c.	

Questions and Conclusions

1. Compare your data with the data of the other groups. From the class data, determine the average values.
2. What can you conclude about the tap water in your area?

CAREER OPPORTUNITY

Plumber

As the name suggests, plumbers first worked with the soft metal lead (Latin name *plumbum*). This metal was used as packing to seal joints in cast iron pipes. Today plumbers work with many new materials. They must have a knowledge of alloys when choosing solders. They must also have a working knowledge of metallurgy, including the properties of copper and solder, as well as metal-cleaning techniques. Hard water causes mineral deposits to build up in plumbing systems, so plumbers also have to be knowledgeable in this area.

Today certain plastics are suitable for many plumbing jobs and the plumber has to keep up to date on new products and the laws that apply to their uses. Plumbers must also have a knowledge of water supply, sanitary waste disposal, and local plumbing codes.

Plumbers learn their trade through an apprenticeship program. Linkage programs with community colleges may be available. If you would like to work in this area, consult your student services department.

Exercises

Key Words

Review

1. Describe the system Berzelius devised for assigning symbols to the elements.
2. List five natural sources of elements.
3. a. What does the word periodic mean?
   b. What are the two names given to rows of elements that are arranged vertically in the periodic table?
   c. What is the name given to rows of elements that are arranged horizontally in the periodic table?
4. How are the elements arranged in the modern periodic table?
5. a. State which of the following elements are nonmetals:

i. **Li**	iii. **Si**	v. **F**	vii. **Ca**
ii. **B**	iv. **C**	vi. **Al**	viii. **P**

   b. Give a use for each of the elements listed in 5a.
6. Name two elements with properties like
   a. tellurium, $z = 52$
   b. cesium, $z = 55$
7. In the periodic table, arsenic, $z = 33$, is surrounded by four other elements: $z = 15$, $z = 32$, $z = 34$, and $z = 51$. Which of these have properties resembling arsenic's?
8. a. Which of the Group 1 elements that you studied is the most reactive?
   b. From your answer to 8a, what can you conclude about reactivity and the size of an atom within a group?
9. State three properties of metals and three properties of nonmetals.
10. Give three examples of alkaline earth metals.
11. a. Predict how barium would react with water and with acid.
    b. Write word equations for your answers to 11a.
12. a. Give two examples of noble gases.
    b. Are noble gases metallic or nonmetallic elements?
13. a. Give two examples of halogens.
    b. Are the halogens metallic or nonmetallic elements?
14. a. Give two examples of alkali metals.
    b. To which group do the alkali metals belong?
15. What is the difference between permanent and temporary water hardness?

16. The compounds listed in the following chart are ingredients found in common household products. Read the labels of products in the kitchen, bathroom, and laundry room and on a chart like the one below, identify products that contain these ingredients.

Compound	Formula	Household Product
acetic acid	CH_3COOH	
acetylsalicylic acid	$C_9H_8O_4$	
aluminum hydroxide	$Al(OH)_3$	
caffeine	$C_8H_{10}O_2N_4$	
isopropyl alcohol	C_3H_8O	
magnesium silicate	$MgSiO_3$	
tin(II) fluoride	SnF_2	
zinc carbonate	$ZnCO_3$	
sodium chloride	$NaCl$	
sucrose	$C_{12}H_{22}O_{11}$	

Extension

1. How do water softeners soften water?
2. Predict some of the chemical and physical properties of element 55 of the periodic table.
3. Research element 106 of the periodic table. State the country in which it was discovered, the scientist who discovered it, and the year it was discovered.
4. What are the roles of the following elements in the human body?

   a. fluorine   c. phosphorus   e. potassium
   b. sodium   d. iron   f. iodine

5. Alzheimer's disease is a brain disease commonly confused with senility. The element aluminum may play a role in this biochemical disorder. Refer to your library resource center and discuss the causes of Alzheimer's disease.

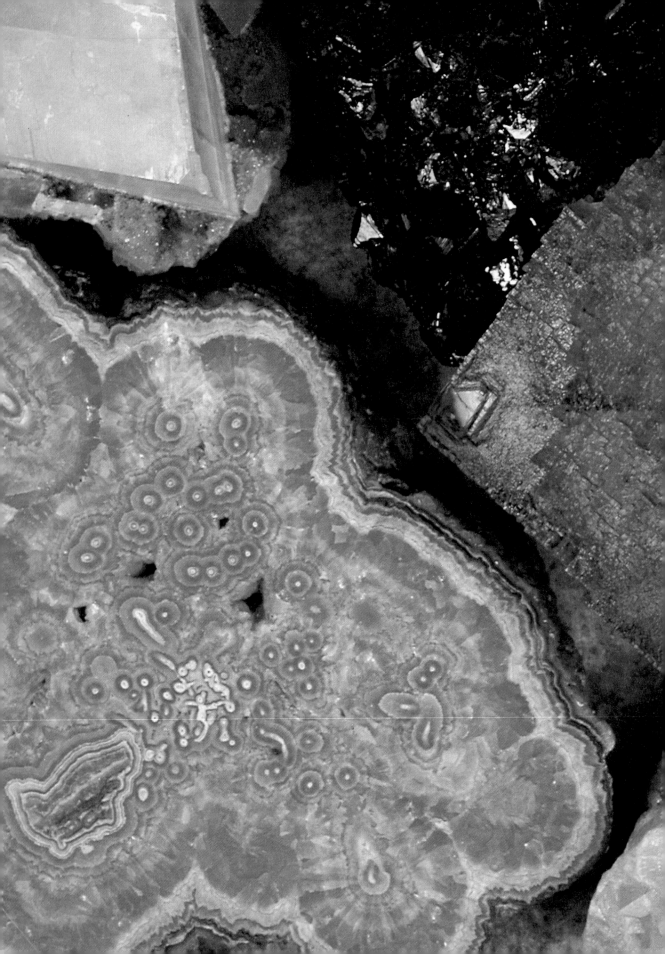

CHAPTER 7
IONIC
BONDING

In Chapter 2 you learned that substances have different properties. These differences allow substances to be separated. This also means that each element has its own unique properties and uses. Two major types of elements, metals and nonmetals, have distinct positions on the periodic table of the elements.

Only a few of the six million known substances on earth exist as elements. Most substances consist of atoms of two or more different elements joined together to form compounds. The atoms in these compounds are held together by forces that scientists call chemical bonds.

Chemists use different kinds of chemical bonds to explain the structure of matter. In this chapter you will learn about one particular type of chemical bond called the ionic bond. You have already learned how ions are formed when a neutral atom gains or loses electrons. These ions can form ionic bonds with each other.

Key Ideas

- Many neutral atoms gain or lose one or more electrons to form ions.
- Elements of Groups 1, 2, and 13 lose electrons to form cations.
- Elements of Groups 16 and 17 gain electrons to form anions.
- Ionic bonds are formed when an element donates one or more electrons to another element.
- Some elements may donate different numbers of electrons.
- Noble gases tend to neither gain nor lose electrons.

7.1
CHEMICAL BONDING

The architecture of a building refers to its size, shape, and the material of which it is built. The architecture of chemical substances refers to the same type of thing: the different types of atoms of which the molecule is composed and how they are put together. Architects use models to illustrate what their project will look like. Scientists use models to illustrate the structure of particles that cannot be seen.

Toronto's SkyDome is an example of unusual architecture. The inset photo shows a model of this building.

Forces of attraction between the atoms in a substance are called chemical bonds. Knowledge of chemical bonds is important in order to understand the chemical reactions that different substances will undergo, as well as their physical properties.

This chapter deals with ionic bonds, which occur because of the attraction of positive and negative ions for each other. Substances that are ionically bonded form ionic solids. These solids consist of positive and negative ions, where positive ions have negative neighbors and negative ions have positive neighbors. The ionic solid is held together by the very strong attraction between oppositely charged ions.

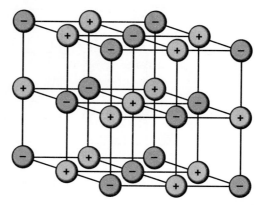

Figure 7.1   An ionic solid

In the following activity you will be doing a series of tests to see whether different substances that are ionically bonded behave in a similar way.

7.2 ACTIVITY
PROPERTIES OF
IONIC COMPOUNDS

All ionic compounds have several properties in common. These properties can be used to determine whether a substance is ionic or not. As you do each test, think of how the results of each test might tell you something about the strength of ionic bonds; that is, how strongly the ions are held together.

Materials

burner	beakers (6)
distilled water	sodium chloride
conductivity apparatus	sodium iodide
scoopula	potassium chloride
deflagrating spoon	potassium iodide
watch glasses (4)	

Figure 7.2 Conductivity apparatus

Testing the conductivity
of a solution

Method

1. Prepare a data table in your notebook. You will be testing four properties of four different ionic compounds.

2. Smell each of the four substances to see if its odor is strong, weak, or not noticeable. Record your results.

3. Use a burner to heat each of the solids in a deflagrating spoon. If the solid melts, record its melting point as *low*. If it does not melt, record its melting point as *high*.

4. Place a few crystals of each solid on watch glasses. Try to crush the crystals by rubbing them between your fingers. Then use a scoopula to try to crush them. Record the results as hard or soft depending on whether or not the crystals can be crushed.

5. Place the electrodes in a beaker of distilled water. Then place the electrodes in a solution of each of the ionic substances. Observe the results.

6. Place the electrodes in a beaker of distilled water, then slowly add sodium chloride to the water. Record whether the solution is a conductor or nonconductor.

Questions and Conclusions

1. a. From your observations, do ionic compounds have high or low melting points?
   b. From your answer to 1a, do you think that ionic bonds are weak or strong?

2. a. Are ionic compounds hard or soft?
   b. From your answer to 2a, can you better decide whether ionic bonds are weak or strong?

3. Do solutions of ionic compounds conduct electricity?

7.3

HOW DOES AN IONIC SOLUTION CONDUCT ELECTRICITY?

In Activity 7.2 you saw that ionic solutions conduct electricity. Solutions of ionic substances conduct electricity because the particles in an ionic substance are charged and because they dissolve to form a solution.

Electricity is the flow of electrons through a wire. When you turn on the lights in your home, electrons flow into the light bulbs through wires that run into your home from the outside.

In ionic solutions, the flow of electricity is caused by charged ions in the solution. Cations, or positive ions, move toward the negative electrode, or cathode. Anions, or negative ions, move toward the positive electrode, or anode. This movement of ions toward the oppositely charged electrode in the solution results in the flow of electric charge.

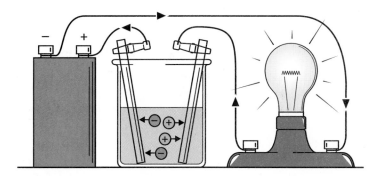

Figure 7.3   The flow of electric charge in an ionic solution

Substances that dissolve to produce conducting solutions are called electrolytes. Electrolytes can either be strong or weak. A strong electrolyte is one that conducts electricity very well. A weak electrolyte is one that conducts electricity poorly.

Try These

1. Why do cations move toward the cathode?
2. What types of substances conduct electricity in solution?
3. What term is used to describe a substance if its solution conducts electricity very well?
4. What would you conclude about a substance if its solution does not conduct electricity at all?

7.4
CATIONS

Ions are formed when neutral atoms gain or lose electrons. Once this happens the atom becomes charged because it has more or fewer electrons than the neutral atom did.

Atoms gain or lose electrons to become more stable. Many atoms are most stable when their outer energy level is filled, or has eight electrons in it. This condition is often referred to as the stable octet.

Certain groups of atoms in the periodic table lose electrons to form ions called cations. When atoms in these groups lose electrons it results in the forming of a stable octet. The following cations each come from a different group. Bohr-Rutherford diagrams are used to show the electron arrangement of atoms and ions.

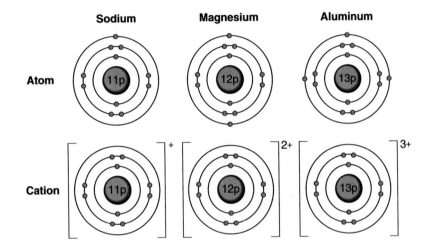

Figure 7.4   Sodium, magnesium, and aluminum atoms and cations

Try These

1. How many electrons does each neutral atom illustrated in Figure 7.4 have in its outer level?

2. Explain why a sodium ion has a charge of 1+; a magnesium ion 2+; an aluminum ion 3+.

3. From what group number in the periodic table does each of the cations in Figure 7.4 come?

4. From your answer to question 3, what could you predict about the number of electrons a potassium atom will lose? a calcium atom? a boron atom?

5. State a general rule for Groups 1, 2, and 13, relating group number to the number of electrons lost.

7.5
ANIONS

Other groups of atoms gain electrons to form ions called anions. These groups gain electrons to become more chemically stable. The anions shown in the Bohr-Rutherford diagrams in Figure 7.5 each come from a different group.

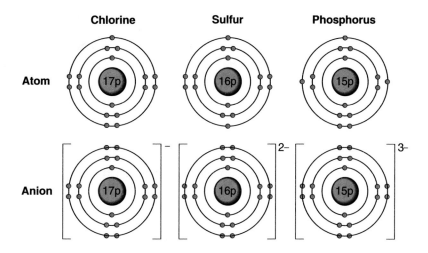

Figure 7.5   Chlorine, sulfur, and phosphorus atoms and anions

Try These

1. How many electrons does each neutral atom of Figure 7.5 have in its outer level?

2. Explain why a chlorine anion has a charge of 1⁻; a sulfur anion 2⁻; a phosphorus anion 3⁻.

3. Name the group number of the periodic table from which each of these anions come.

4. From your answer to question 3, what can you predict about the number of electrons a fluorine atom will gain? an oxygen atom? a nitrogen atom?

5. State a general rule for these three groups of the periodic table, relating group number to the number of electrons gained.

As you have observed, it is easier for some groups to lose electrons. For other groups it is easier to gain electrons. These two statements bring a number of questions to mind. For instance, why do certain groups gain electrons while other groups lose electrons to form a stable octet? What causes this gain or loss of electrons? Can we always predict accurately which groups and which members of a group will be most likely to gain or lose electrons?

The most important factor needed to answer these questions is the atomic radius of an atom. The atomic radius is the estimated distance from the center of the nucleus of the atom to the outer edge of the atom.

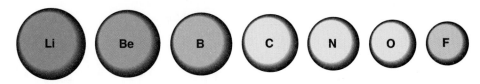

Figure 7.6   The sizes of atoms within a period of the periodic table

Related to atomic radius is ionization energy. The ionization energy of an element is the amount of energy needed to remove an electron from an atom. It is measured in kilojoules per mole (kJ/mol).

7.6
REPRESENTING
ELECTRONS

One way of representing an atom and its electrons is with a Bohr-Rutherford diagram. A Bohr-Rutherford diagram shows all of an atom's electrons.

In 1916 the American chemist G.N. Lewis suggested another method of representing atoms and ions—using dots to represent the electrons in the outermost shell of the atom. The outer-shell electrons are called valence electrons and the outer shell of electrons is called the valence shell. The dots representing electrons are placed around the symbol, up to a maximum of eight dots. The following are some examples of Lewis diagrams, or electron dot diagrams, for some elements.

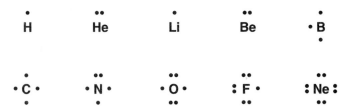

Figure 7.7   G.N. Lewis proposed a simple way of illustrating the valence shell electrons of atoms.

Only the electrons of the outer energy level are shown because these electrons are most responsible for an atom's chemical properties.

For ions, electrons are either added to or removed from the original electron dot diagram. For example, the oxygen atom and anion are written as follows:

The oxygen atom   **The oxygen anion**

The sodium atom and cation are written as follows:

Na   Na⁺

The sodium atom   **The sodium cation**

Try These

1. Draw Lewis diagrams (electron dot diagrams) for
   a. **Na**   b. **Al**   c. **P**   d. **Cl**   e. **Ar**   f. **Ca**

2. Draw Lewis diagrams (electron dot diagrams) for
   a. **Cl⁻**   c. **N³⁻**   e. **Ca²⁺**   g. **Al³⁺**   i. **P³⁻**
   b. **Mg²⁺**   d. **K⁺**   f. **F⁻**   h. **Ne**   j. **S²⁻**

7.7
IONIC BONDING

Elements on the left side of the periodic table tend to form cations by losing one or more electrons. Elements on the right side of the periodic table tend to form anions by gaining one or more electrons.

Sodium is a highly reactive soft silvery metal. Each sodium atom has one outer electron. If it can lose this electron it will have a filled outer energy level.

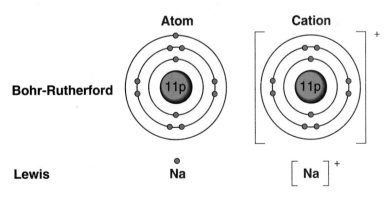

Figure 7.8   The Bohr-Rutherford and Lewis diagrams for a sodium atom and cation

Chlorine is a yellow-green poisonous gas. Each chlorine atom has seven outer electrons. If it can gain one electron it will have a stable octet.

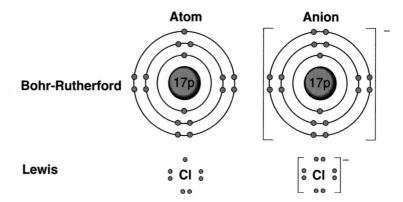

Figure 7.9   The Bohr-Rutherford and Lewis diagrams for the chlorine atom and anion

If a sodium atom collides with a chlorine atom, something similar to a tug-of-war for sodium's outer electron occurs. This outer electron is pulled in two different directions at once. It is pulled toward its own nucleus, and it is pulled toward chlorine's nucleus.

Figure 7.10   The tug-of-war for sodium's single valence electron

Since chlorine has the greater attraction for electrons, it wins the tug-of-war. Chlorine fills its outer energy level, forming a stable octet, and becomes an anion whose charge is 1-. By losing its outer electron, sodium is left with a stable octet and becomes a cation whose charge is 1+. Figure 7.11 illustrates the result of the reaction between sodium and chlorine.

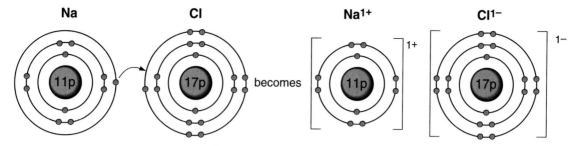

Figure 7.11 A sodium atom and chlorine atom combine to form sodium chloride.

Chlorine, a green gas, and sodium, a grey metal, combine to form sodium chloride, a white crystal.

The oppositely charged ions of sodium and chlorine attract each other very strongly. This attraction causes an ionic bond to form. The result of this ionic bond is the ionic compound sodium chloride. Note that the net charge of an ionic compound is always zero. In the case of sodium chloride it is $(1+) + (1-) = 0$.

Sodium chloride is made up of small, white, cubic crystals that dissolve readily in water and give food more flavor. Interestingly, sodium chloride is completely unlike either of the two elements that combine to form it. Sodium is a poisonous metal and chlorine is a poisonous gas.

Another example involving magnesium and chlorine is illustrated in Figure 7.12. In this case magnesium loses two valence electrons to become a magnesium cation with a charge of $2+$ and combines with two chloride anions, each with a charge of $1-$. Note that the net charge for the compound is zero.

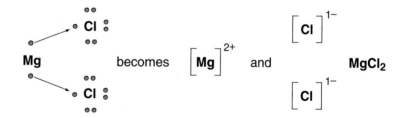

Figure 7.12 A magnesium atom and two chlorine atoms combine to form magnesium chloride.

Try These

1. Which noble gas has the same number of electrons as a sodium ion?

2. Which noble gas has the same number of electrons as an ion of chlorine?

3. a. Use Lewis diagrams to describe the formation of the ionic compound lithium fluoride.

   b. Which noble gas has the same number of electrons as an ion of fluorine?

4. List four ions with the same number of electrons as neon.

5. Use Lewis diagrams (electron dot diagrams) to show the ionic bond formation in each of the following pairs of elements:

   a. **Li Cl**   d. **Al Cl**   g. **Ba S**
   b. **K S**   e. **Mg Cl**   h. **Na O**
   c. **Ca Cl**   f. **Mg N**

7.8 ACTIVITY
FORMATION OF AN
IONIC COMPOUND

In Section 7.7 you learned that metals lose electrons to form cations. Nonmetals gain electrons to form anions. The attraction between oppositely charged ions leads to the formation of an ionic bond and an ionic compound.

In Part A you will mix two solutions containing ions and observe the mixture for evidence that an ionic compound has formed.

In Part B you will react a metal with an acid to produce an ionic compound and a gas. Because the ionic compound is soluble, water must be evaporated to recover the ionic substance.

Materials

large test tubes (3)   evaporating dish
silver nitrate solution   burner
sodium chloride solution   retort stand
mossy zinc   ring clamp
hydrochloric acid

Caution: Silver nitrate and hydrochloric acid are corrosive and cause burns. Handle them with care. Wear goggles and an apron. If you spill any on yourself, rinse it off immediately with large amounts of cold water and have your partner tell your teacher.

Method

Part A

1. Pour 5 mL of silver nitrate solution into a test tube.
2. Pour 5 mL of sodium chloride solution into the second test tube.
3. Add the sodium chloride solution to the silver nitrate solution in the large test tube.
4. Observe the mixture and record your observations.

Part B

1. Half-fill a test tube with hydrochloric acid.
2. Add a piece of mossy zinc to the test tube.
3. Record your observations.
4. After a few minutes pour some of the solution from the test tube into an evaporating dish.
5. Evaporate the liquid by heating it over a burner.
6. Examine the residue.

Questions and Conclusions

Part A

1. What was the color of the solid that formed?
2. Ions were present in both solutions. What type of solid compound do you think was formed?

Part B

1. What evidence indicated that a reaction had occurred?
2. What evidence was there that an ionic solid had formed?

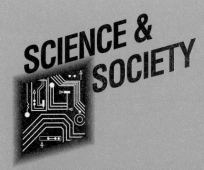

SCIENCE & SOCIETY

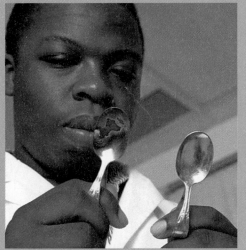

Tarnished and polished silverware

Removing the Tarnish From Silver

Silverware tarnishes when a chemical reaction occurs between the silver and sulfur in the air or in food. The black "tarnish" is the ionic compound silver sulfide.

Tarnishing can be controlled by proper storage of silver. Tarnished silver can be restored by polishing it with cleansers that remove the silver sulfide.

7.9

NAMING BINARY COMPOUNDS

In Activity 7.8 silver ions reacted with chloride ions to produce an ionic compound called silver chloride.

$$\text{silver} + \text{chlorine} \longrightarrow \text{silver chloride}$$

Compounds such as silver chloride that contain only two elements are known as binary compounds. Many ionic substances are binary compounds. You have already seen that sodium and chlorine react to form the ionic compound sodium chloride, and magnesium and chlorine react to form magnesium chloride. Binary compounds are not difficult to name. There are only two rules.

1. Write the cation first, using the name of the element.
2. Write the anion second, dropping the usual ending **-ine, -ium, -ogen,** etc. and replacing it with **-ide.**
   For example, chlorine becomes chloride.

Table 7.1   The names of some common anions

Element	Anion
fluorine	fluoride
chlorine	chloride
bromine	bromide
iodine	iodide
oxygen	oxide
sulfur	sulfide
nitrogen	nitride
phosphorus	phosphide

The name of a binary compound is a two word name: the name of the cation followed by the name of the anion written with the **-ide** ending. For example the reaction of potassium and iodine is as follows:

$$\text{K}^+ + \text{I}^- \longrightarrow \text{KI}$$

The ionic compound formed is named potassium iod**ide**.

Some binary compounds, such as silver sulfide **(Ag$_2$S)** and magnesium chloride **(MgCl$_2$),** have two or more of one of the ions in the formula. This does not change the way the compound is named.

Try These

1. Name the following compounds:

   a. **MgO**   c. **NaBr**   e. **Na$_2$O**

   b. **LiF**   d. **CaO**   f. **MgF$_2$**

2. Prepare a chart like the one shown and fill in the missing ionic compound names. The first one has been done for you.

Formula	Name
CaBr$_2$	calcium bromide
AlN	
NaI	
Ag$_3$P	
BaCl$_2$	
AlCl$_3$	
K$_2$S	
Na$_2$O	

7.10

HOW TO WRITE IONIC FORMULAS

Cations of Groups 1, 2, and 13 lose one, two, or three electrons and have charges of 1+, 2+, or 3+.

Group 1	Group 2	Group 13
Li$^+$	Be^{2+}	B^{3+}
Na$^+$	Mg^{2+}	Al^{3+}
K$^+$	Ca^{2+}	

Anions of Groups 15, 16, and 17 gain three, two, or one electrons and have charges of 3-, 2- or 1-.

Group 15	Group 16	Group 17
N^{3-}	O^{2-}	F$^-$
P^{3-}	S^{2-}	Cl$^-$

Ions always combine to give a total electrical charge of zero in an ionic compound. Therefore, in the case of Group 1 and Group 17 ions, one Group 1 and one Group 17 ion will combine to form a neutral ionic compound.

Try These

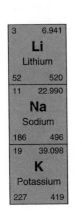

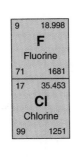

1. Copy the chart below and write in the formulas for the ionic compounds formed by combining Groups 1 and 17. One has been done for you.

	F$^-$	Cl$^-$
Li$^+$		LiCl
Na$^+$		
K$^+$		

2. Write the names of the six ionic compounds formed in question 1.

The Trend Continues

Group 2 elements lose two electrons and Group 16 elements gain two electrons. Therefore, one Group 2 ion and one Group 16 ion will combine to form neutral ionic compounds in the same way Group 1 and 17 did. This trend also continues for one Group 3 and one Group 15 ion combining to form an ionic compound.

Group 2	Group 16
Be^{2+}	O^{2-}
Mg^{2+}	S^{2-}
Ca^{2+}	

Try These

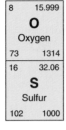

1. Copy the chart below and complete it by writing the formula of the ionic compounds formed by Groups 2 and 16. The first one has been done for you.

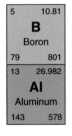

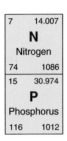

	O^{2-}	S^{2-}
Be^{2+}	BeO	
Mg^{2+}		
Ca^{2+}		

2. Name the six ionic compounds formed in question 1.
3. Copy the chart below and complete it by writing the formulas for the four ionic compounds formed when Groups 13 and 15 combine.

	N^{3-}	P^{3-}
B^{3+}		
Al^{3+}		

4. Name the four ionic compounds formed in question 3.

The Cross-over Rule

There are other possibilities for ionic bonding aside from those mentioned so far. Many types of ionic compounds are formed between groups that do not combine in a one-to-one ratio.

For example, lithium will combine with oxygen, but oxygen needs two electrons to fill its outer energy level. It must take the electrons from two lithium atoms since each lithium atom has only one electron to donate.

The ionic formula of lithium oxide is **Li_2O**. This means that there are two ions of lithium to one ion of oxygen in the compound. This is shown by writing the subscript (something written below the line) immediately after the atom to which it refers.

A simple way to figure out the ionic formula is called the cross-over rule. This rule consists of the following steps:

Step 1

Write the symbols and charges of the elements in the compound, metal first, nonmetal second.

$Al^{3+}S^{2-}$ $Mg^{2+}Cl^{1-}$ $Mg^{2+}O^{2-}$

Step 2

Cross over the number of the charges for each ion and leave out the charge sign.

$Al^{3+}S^{2-}$ $Mg^{2+}Cl^{1-}$ $Mg^{2+}O^{2-}$

Al^2S^3 Mg^1Cl^2 Mg^2O^2

Step 3

Determine the highest common factor for the crossed-over charges.

1   1   2

Step 4

Divide the crossed-over charges by the highest common factor.

Al^2S^3 Mg^1Cl^2 Mg^2O^2

Step 5

Write the resulting figures as subscripts after the ion they refer to. The number 1 is always understood to be there and so it may be omitted from the formula.

Al_2S_3 $MgCl_2$ MgO

Try These

1. Write the formulas for the following binary compounds:

   a. potassium chloride   e. calcium oxide
   b. sodium sulfide   f. magnesium fluoride
   c. aluminum sulfide   g. potassium phosphide
   d. calcium chloride   h. silver oxide

7.11

NAMING BINARY COMPOUNDS FOR ELEMENTS THAT HAVE MORE THAN ONE CHARGE

Sometimes elements can have more than one charge. When this happens it is possible for that element to form more than one compound with another element. Most of the transition metals found between Groups 2 and 13 of the periodic table are multivalent, i.e., they can have more than one valence or charge.

An example of this is the transition metal copper. Copper can lose either one or two electrons to form cations with a charge of either 1+ or 2+. Copper can combine with oxygen to form two different oxides, Cu_2O, which is red, and CuO, which is black.

Two oxides of copper: Cu_2O (left), CuO (right)

These compounds are named using the Stock system. They are named like any other binary compound. The only difference is that the charge of the particular ion is written in Roman numerals, in parenthesis, after the element to which it refers. For example, the two oxides of copper are as follows.

Formula	Charge on Copper	Name
Cu_2O	1+	copper(I) oxide
CuO	2+	copper(II) oxide

This system was developed by the German chemist Alfred Stock (1876–1946). He was one of the first scientists to give warnings about the dangers of mercury poisoning.

Table 7.2   Charges of some transition metals

Element	Symbol	Charges
copper	Cu	1+, 2+
iron	Fe	2+, 3+
mercury	Hg	1+, 2+
tin	Sn	2+, 4+
manganese	Mn	2+, 4+, 7+
cobalt	Co	2+, 4+
gold	Au	1+, 3+
lead	Pb	2+, 4+

Try These

1. Write the formulas for the following binary compounds:
   a. iron(II) oxide
   b. iron(II) chloride
   c. cobalt(II) phosphide
   d. lead(IV) chloride
   e. lead(II) chloride
   f. mercury(I) oxide
   g. gold(III) fluoride
   h. tin(II) fluoride

2. Write the names for the following compounds:
   a. CuS c. Cu_3N e. $FeBr_3$ g. PbO_2
   b. $CoCl_2$ d. SnF_4 f. Ni_2S_3 h. Hg_2S

7.12 ACTIVITY
THE STRUCTURE OF IONIC CRYSTALS

Many ionic compounds have unique shapes. This shape is referred to as its crystalline structure. In this activity you will examine some crystals under the compound light microscope. The microscope will help you see that each ionic solid has its own crystalline structure.

Materials

compound light microscope
microscope slides (4)
chrome alum crystals
sodium tartrate crystals
sodium chloride crystals
sodium thiosulfate crystals

Method

1. Carefully place one or two crystals of each of the compounds on separate microscope slides.
2. Examine the crystals under the microscope's low and medium power. You will need to adjust the diaphragm of the microscope to obtain the best lighting.
3. Sketch each kind of crystal.

Questions and Conclusions

1. Describe the shape or structure of each of the crystals.
2. What is the best light for viewing crystals with a microscope? Why?

Amethyst crystals inside a geode

Postlab Discussion

The growth of crystals is caused by small units of the crystal adding onto a larger unit, the seed crystal. For example, the reaction between ions of chlorine and sodium does not stop with the formation of a single ion pair. More ion pairs are added on and the crystal builds up into a crystal lattice, with each sodium ion being surrounded by six chloride ions. A crystal lattice is the three-dimensional shape formed by linking ion pairs.

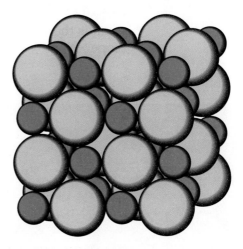

Figure 7.13   A sodium chloride lattice. Compare this figure to Figure 7.1, page 125.

The crystal lattice grows the same way for any ionic substance. Crystals of a particular substance always have a definite shape. For many ionic compounds, the shape is cubic.

7.13 ACTIVITY
CLEAVAGE
OF CRYSTALS

In this activity you will cut large crystals into smaller ones and find out the shape of the smaller crystals.

Materials

rock salt or other large crystal
single-edged razor blades

Caution: Use care when handling razor blades.

Method

1. Your teacher will provide you with some crystals. Observe them carefully and make notes on the size and the shape of the crystals.
2. Use a single-edged razor blade to carefully make a cut through one of the crystals, parallel to one of its sides. This is called a cleavage plane.
3. Observe the cut-off crystals carefully.
4. If your crystal is large enough, repeat step 2.

The regular shape of a naturally formed sodium chloride (halite) crystal

Questions and Conclusions

1. Describe the size and shape of your original crystal.
2. Describe the size and shape of the crystals after cleavage.

7.14
GEMS

Diamonds, rubies, emeralds, and sapphires are very special crystals. Because of their great beauty and rarity, they are called precious stones. Many others like jade, garnet, zircon, aquamarine, amethyst, and peridot are also very beautiful but are more plentiful than precious stones. These are called semi-precious stones.

With the exception of a few gems made of organic materials like pearls and opals, most **gems** are minerals and have crystalline structures. The photograph below is of rubies, which are made of the compound aluminum oxide. These gems will cleave along certain lines. Therefore, jewelers must be very careful when they cut these valuable stones.

The material that makes up both sapphires and rubies is the binary compound aluminum oxide, **Al_2O_3**. The red color of the ruby is caused by a small amount of chromium impurity.

Synthetic gems, although often beautiful, are far less valuable than a real gem. Some synthetic gems are of such high quality that they might easily be mistaken for a real gem. For example, cubic zirconia stones look very much like diamonds. Gem experts use sophisticated instruments such as spectroscopes or X-ray diffraction analysis to detect synthetic and real gems. They may even be called on to help in crime prevention if a synthetic gem is fraudulently sold as a real gem.

The diamonds on the left will be cut to produce gems like the ones on the right.

Small pieces of diamond, the hardest stone known, are used as drill bits and saw blades. Because of its hardness, a diamond bit can cut through most materials with ease.

A diamond bit drill (left) and saw blade (right)

SCIENCE & TECHNOLOGY

Liquid Crystals

Not all crystals are solid. Liquid crystals and their properties have interested scientists for nearly a century. Because of their unique properties, chemists have found some very practical applications for liquid crystals.

In its liquid crystal form, the substance can transmit and reflect light like a solid crystal. There are three types of liquid crystals. Each has a different molecular arrangement.

One type, nematic liquid crystals, consist of tube-shaped molecules. The forces between the molecules are strong enough that they move as a group. They also respond to an electric field. This is the type of crystal used in liquid crystal displays (LCD).

The LCD is made up of small glass tubes containing nematic liquid crystals. When an electric charge is present, the crystals rotate in the direction of the field. Light is blocked in certain areas to form a displayed pattern.

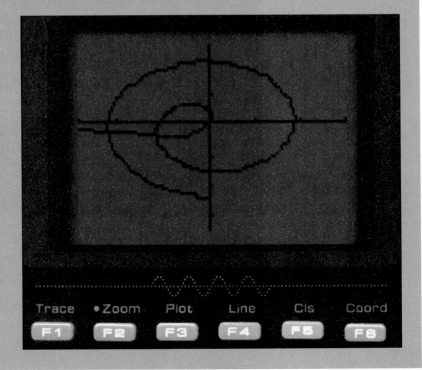

CAREER OPPORTUNITY

Gemologist

Gems are called precious stones because of their rarity, beauty, or other unique qualities. A gemologist is an expert on gems.

Gemologists must know how to judge and appraise the quality and value of many different kinds of stones. They must also know the characteristics of gems that add to their value as precious stones. To do this, the gemologist must be familiar with the chemical composition of gems as well as any historical or mythological information that is connected to a particular stone.

Key Words

Exercises

Review

1. State the charge sign and number for the ions of each of the following atoms:

   a. **Mg**   c. **S**   e. **Cl**   g. **B**

   b. **K**   d. **Al**   f. **O**   h. **P**

2. Which ions have the following atomic structures?

	Number of Protons	Number of Electrons
a.	3	2
b.	19	18
c.	8	10

3. Why are cations positively charged?
4. Give three examples of ionic solids.
5. Briefly explain how an ionic solution conducts electricity.
6. What is a stable octet and why is it so important?
7. Which of these atoms has the largest atomic radius?

   i. a. **Li**   b. **Na**   c. **K**   d. **Rb**   e. **Cs**

   ii. a. **Li**   b. **Be**   c. **B**   d. N   e. **F**

8. Why does aluminum have a lower ionization energy than magnesium?
9. Why does calcium have a lower ionization energy than beryllium?
10. Draw electron dot diagrams for each of the following:

    a. atoms: i. **Mg**   ii. **K**   iii. **He**

    b. ions: i. **Na$^+$**   ii. **Be^{2+}**   iii. **F$^-$**

    c. Name the noble gas with the same number of electrons as each of the ions in 10b.

11. Name the following ionic compounds:

    a. **NaI**   c. **Li$_2$O**   e. **KI**   g. **CaO**   i. **AuCl$_3$**

    b. **CaCl$_2$**   d. **AlF$_3$**   f. **Mg$_3$P$_2$**   h. **Cu$_2$O**   j. **SnCl$_4$**

12. Write the formulas for the compounds formed by the following and name them:

    a. potassium and iodine

    b. barium and chlorine

    c. sodium and oxygen

    d. aluminum and nitrogen

    e. lithium and sulfur

    f. aluminum and oxygen

13. Write the formulas for each of the following compounds:
   a. copper(II) fluoride
   b. mercury(I) oxide
   c. tin(II) sulfide
   d. lead(IV) oxide
   e. manganese(VII) oxide
   f. cobalt(II) sulfide

14. a. Copy the chart below and write in the ionic formulas to complete it. A few have been done as examples for you.

	F^-	Cl^-	O^{2-}	S^{2-}	N^{3-}	P^{3-}
H^+	HF					
Li^+						
Na^+						
K^+						
Be^{2+}			BeO			
Mg^{2+}						
Ca^{2+}		$CaCl_2$				

   b. Name the ionic compounds represented on the chart in 14a.

Extension

1. List the names of gems that you know. Use reference materials to find out the chemical composition of each type of stone.

2. Visit a jewelry store in your area. Ask the jeweler to show you the diamond scope. You may be allowed to view a diamond through this microscope. Ask for an explanation of the factors that affect the value of diamonds. Ask about the rating scale that is used to classify diamonds. Present your results to the class.

3. Build large scale models of some ionic crystals. Use foamed polystyrene spheres and toothpicks or pipe cleaners. These can be painted and used to decorate your lab or classroom.

4. People often feel that they are affected by atmospheric conditions that lead to a build-up of charge in the air. Research the use of commercial negative ion generators. Present your results to the class.

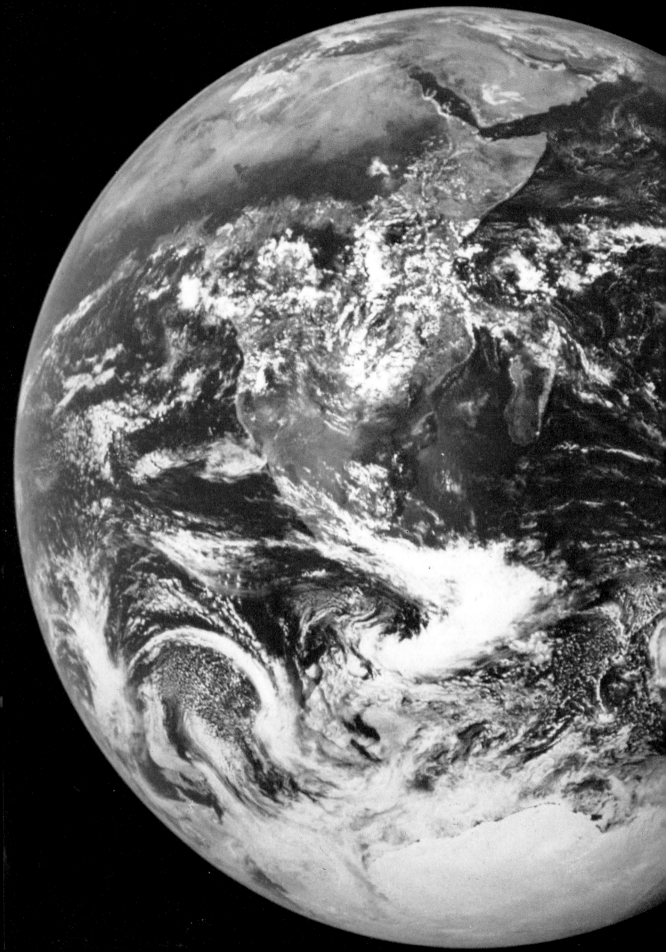

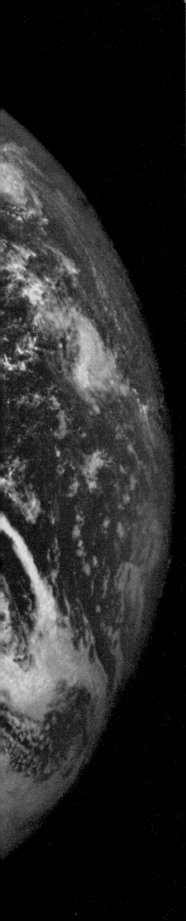

CHAPTER 8
COVALENT
BONDING

In Chapter 7 you learned that metallic elements tend to lose electrons to form cations, while non-metallic elements tend to gain electrons to form anions. Since oppositely charged particles attract each other, many metals form ionic bonds with nonmetals.

However, many nonmetals form compounds with other nonmetals. How can this be? Since nonmetals tend to gain electrons, they should repel each other, shouldn't they?

This chapter will answer that question by intro-ducing you to the covalent bond. This type of bond occurs between two or more nonmetals.

Key Ideas

- Two atoms can share electrons to form covalent bonds.
- Equal sharing of electrons results in nonpolar covalent bonds.
- Unequal sharing of electrons results in polar covalent bonds.
- Atoms can share two, four, or six electrons to form single, double, or triple covalent bonds respectively.
- Covalent compounds have a distinct set of properties that distinguish them from ionic compounds.
- Structural formulas can be used to represent covalent compounds.
- Polar solutes dissolve in polar solvents. Nonpolar solutes dissolve in nonpolar solvents.
- Liquid mixtures can be miscible or immiscible.

The earth's atmosphere contains many gases in the form of covalent molecules, including nitrogen (N_2), oxygen (O_2), water (H_2O), carbon dioxide (CO_2), and ozone (O_3).

8.1
COVALENT BONDING

Covalent bonding happens when the nuclei of two atoms each try to gain electrons. Neither nucleus is able to completely pull an electron away from the other atom's nucleus. As a result, the atoms must share the electrons between them. Covalent bonding does not involve the transfer of electrons as ionic bonding does.

Covalent bonding often occurs between two atoms of the same element. An example of this is the formation of hydrogen gas. This happens because each hydrogen atom needs an additional electron to fill its outer energy level. Since neither can remove the other's electron, they have to make a compromise in order to fill their outer energy levels. They do this by sharing the outer electrons between them. A covalent bond is formed. Both hydrogen atoms, in effect, have two electrons in their outer energy levels. The result is a molecule of hydrogen gas, (H_2), which is more stable than either of the two lone atoms of hydrogen.

The electron dot diagram below shows the formation of a molecule of hydrogen:

Hydrogen molecule

Oxygen molecule

Nitrogen molecule

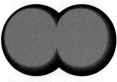

Chlorine molecule

Figure 8.1   The diatomic molecules of some gases

$$H \circ \; + \; H \circ \longrightarrow H \circ\!\!\circ H$$

Chlorine in Group 17 has seven outer electrons. By sharing electrons, each of two chlorine atoms can obtain a stable octet, as shown in the electron dot diagram below.

$$\circ\!\!\circ \quad \circ\!\!\circ$$
$$\circ\; Cl \circ\!\!\circ Cl \;\circ$$
$$\circ\!\!\circ \quad \circ\!\!\circ$$

Elemental gases such as hydrogen and chlorine almost always exist as molecules made up of two atoms of the same element. These types of molecules are called diatomic molecules. The prefix **di** means two.

Different types of atoms can also combine to form covalent bonds and covalent molecules. For example, water (H_2O) is composed of two hydrogen atoms and one oxygen atom.

$$\circ\!\!\circ \qquad\qquad\qquad \circ\!\!\circ$$
$$\circ\; O \circ\; + \; H \circ \; + \; H \circ \longrightarrow H \circ\!\!\circ O \circ\!\!\circ H$$
$$\circ\!\!\circ \qquad\qquad\qquad \circ\!\!\circ$$

Oxygen in Group 16 needs to share two electrons. Hydrogen needs one electron. Each oxygen atom combines with two hydrogen atoms.

Try These

1. Draw an electron dot diagram for fluorine gas (**F₂**).
2. Draw an electron dot diagram for the hydrogen chloride molecule (**HCl**).
3. Draw an electron dot diagram for methane (**CH₄**).
4. Write the formula for the following covalent compounds:

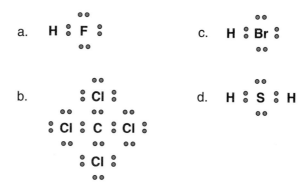

8.2
COVALENT BONDS AND ELECTRO-NEGATIVITY

In diatomic molecules such as **H₂**, **Cl₂**, and **F₂**, the two atoms are identical. Their ability to attract electrons is also identical. Therefore, the electrons are shared equally between the nuclei of the two identical atoms in a diatomic molecule. This type of covalent bond is called a nonpolar covalent bond. Diatomic molecules formed with nonpolar bonds are nonpolar molecules

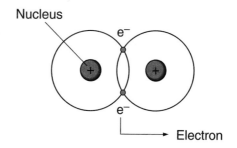

Figure 8.2   Equal sharing of electrons in a hydrogen molecule, **H₂**

In many other covalent compounds, one element has a greater electron-attracting ability than the other(s). The element with the greater attraction pulls the electrons closer to its nucleus. This type of covalent bond is called a polar covalent bond. Since more electrons tend to be in the region of one nucleus rather than the other, the end of the molecule with more electrons concentrated in that region tends to become more negatively charged.

This uneven distribution of charge results in a positive and a negative end to the molecule and results in a polar molecule. Figure 8.3 shows the electron-attracting ability, or electronegativity (blue), of the first 20 elements. The higher the value, the greater the electron-attracting ability.

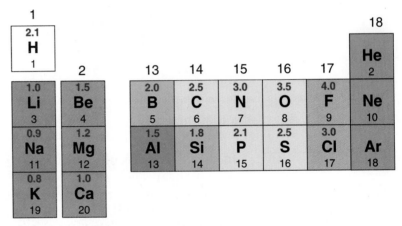

Figure 8.3 Trends in electronegativity for the first 20 elements

Figure 8.3 can be used to find out what type of bond has been formed in any compound. To do this, the difference in electronegativity between the elements in the compound is calculated. A difference between 0 and 0.3 represents equal or almost equal sharing. A nonpolar covalent bond is formed. If the difference is between 0.3 and 1.7, a polar covalent bond is formed. If the difference is greater than 1.7, then the electrons are so unevenly shared that an ionic bond is formed.

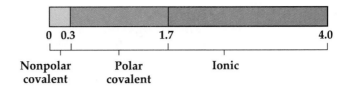

0 0.3	1.7	4.0
Nonpolar covalent	Polar covalent	Ionic

Table 8.1   Types of bonds

Molecule	Calculate Difference in Electronegativity	Type of Bond
$\overset{\bullet\bullet}{\underset{\bullet\bullet}{:}} Cl \overset{\bullet\bullet}{\underset{\bullet\bullet}{:}} Cl :$	Cl = 3.0 Cl = 3.0 ——— 0.0	nonpolar covalent
$H : \overset{\bullet\bullet}{\underset{\bullet\bullet}{N}} : H$ H	N = 3.0 H = 2.1 ——— 0.9	polar covalent
NaCl	Cl = 3.0 Na = 0.9 ——— 2.1	ionic

Try These

1. State whether the following examples are examples of nonpolar covalent, polar covalent, or ionic bonding.
   a. LiF   b. H_2O c. F_2 d. K_2O e. KCl
2. Name the compounds in question 1.
3. State which end of the molecule is negatively charged and which end is positively charged for the polar compounds presented in question 1.

8.3 ACTIVITY
BUILDING
COVALENT
MOLECULES

There are many ways of representing the structure of chemical substances. In this activity you will use ring-shaped candies, gumdrops, and toothpicks to build molecular models. Each gumdrop represents an atom. Each toothpick represents a covalent bond or shared pair of electrons. Ring-shaped candies will be used to represent the shared pairs of electrons.

Materials
toothpicks
ring-shaped candies
gumdrops

Method

1. Decide on a different color of gumdrop for each different type of atom.
2. Build molecular models for the following molecules:
   a. hydrogen gas   d. fluorine gas
   b. chlorine gas   e. hydrogen fluoride
   c. hydrogen chloride   f. water
3. If the covalent bond is polar, use a ring-shaped candy around the toothpick between the atoms to indicate the relative position of the shared electrons in the bond.

Questions and Conclusions

1. Classify the compounds that you constructed as containing polar or nonpolar covalent bonds.
2. In polar covalent bonds, why are the electrons unevenly shared?
3. In the polar molecules formed, which end is slightly positively charged? Which end is negatively charged?

8.4
MULTIPLE BONDS

Elements that share a pair of electrons form single bonds. Elements that share more than one pair of electrons form multiple bonds. Group 16 and 15 elements need more than one electron in their outer shell to form a stable octet. Elements in these groups often form multiple bonds.

Table 8.2   Multiple bonds

Electron Dot Diagram	$\overset{\bullet\bullet}{\underset{\bullet\bullet}{O}} :: \overset{\bullet\bullet}{\underset{\bullet\bullet}{O}}$	$\overset{\bullet\bullet}{N} \vdots \vdots \overset{\bullet\bullet}{N}$
Group number	16	15
Number of electrons needed for each stable octet	2	3
Number of shared pairs	2	3
Type of bond	double	triple

Try These

1. How many pairs of shared electrons are there in a double bond?
2. How many pairs of shared electrons are there in a triple bond?
3. In the molecule $\overset{\bullet\bullet}{O}::C::\overset{\bullet\bullet}{O}$ (CO_2), what kinds of bonds are formed between each oxygen atom and carbon?

8.5
LEWIS STRUCTURES AND STRUCTURAL FORMULAS

Lewis structures are similar to electron dot diagrams. The only difference is that the shared pair or pairs of electrons are represented by lines instead of dots. The electrons that are not involved in the bonding pairs are still represented by dots. The following are some examples of Lewis structures. Each line represents two electrons or a bonding pair.

(Lewis structure)   **H – H** $\ddot{\mathbf{O}} = \ddot{\mathbf{O}}$ $\ddot{\mathbf{N}} \equiv \ddot{\mathbf{N}}$

(Molecular formula)   **H₂**   **O₂**   **N₂**

Lewis structures include electrons not involved in bonding. A simpler method of representing covalent molecules involves only the pairs of electrons that form covalent bonds. These are called structural formulas. In structural formulas a line is used to represent the shared pair of electrons that forms a bond. Unshared electrons are not shown.

Table 8.3   The molecular formula, Lewis structure, and structural formula for some covalent compounds.

Molecular Formula	Lewis Structure	Structural Formula
H_2	H – H	H – H
O_2	$\ddot{O} = \ddot{O}$	O = O
N_2	$\ddot{N} \equiv \ddot{N}$	N ≡ N
CO_2	$\ddot{O} = C = \ddot{O}$	O = C = O
CCl_4	Cl — C — Cl structure with four Cl	Cl, Cl, C, Cl, Cl
NH_3	H — N — H with H above	H — N — H with H above

Lewis structures and structural formulas are useful because they show
- the kinds of atoms in a molecule
- the number of each kind of atom
- the arrangement of the atoms
- the covalent bonds that are present

One disadvantage of Lewis structures and structural formulas is that they are two-dimensional. Molecules exist in three-dimensional space. For example, CCl_4 has a tetrahedral shape and NH_3 is shaped like a triangular pyramid.

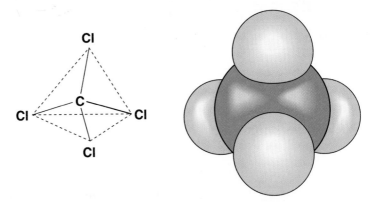

In the carbon tetrachloride molecule, **CCl₄**, a carbon atom is covalently and symmetrically bonded to four atoms.

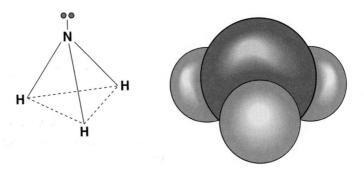

The pyramidal shape of the ammonia molecule, **NH₃**, is the result of no atom being bonded to the lone pair of electrons.

Figure 8.4   Structural formula and shapes of **CCl₄** and **NH₃**

The shape of a molecule often affects its properties. Both CCl_4 and NH_3 contain polar covalent bonds. Because of its shape, CCl_4 turns out to be a nonpolar molecule. The polar bonds are symmetrical and cancel each other's effects. The NH_3 molecule is not symmetrical, so NH_3 is a polar covalent compound. The shape of a molecule as well as the type of bonding will affect the properties of the substance.

Try These

1. Draw Lewis structures and structural formulas for the following molecules:

   a. Cl_2 b. HCl c. BH_3 d. CH_4 e. C_2H_2 f. CH_2O

8.6 ACTIVITY
MORE GUMDROPS
AND TOOTHPICKS

In Activity 8.3 you used toothpicks to represent single covalent bonds between atoms. Each kind of atom was represented by a different color of gumdrop. Compounds containing double and triple bonds can also be represented using these materials. How many toothpicks will you need to represent a double bond? How many for a triple bond?

Materials
toothpicks
gumdrops

Method

1. Decide on a different color of gumdrop to represent each type of atom.
2. Build molecular models for each of the following. Note that a molecule may contain more than one type of bond.
   - a. H_2
   - b. O_2
   - c. N_2
   - d. F_2
   - e. CO_2
   - f. HF
   - g. C_2H_2
   - h. H_2S

Questions and Conclusions

1. Which of the above molecules contain double bonds?
2. Which of the molecules contain triple bonds?
3. Which of the molecules contain more than one type of bond?
4. How many shared pairs of electrons are present in a double bond?
5. How many shared pairs of electrons are present in a triple bond?
6. Which molecules are nonpolar?
7. Which molecules are polar?
8. Which molecule is the most polar?
9. For the polar molecules, state which end of the molecule is positive and which end is negative.

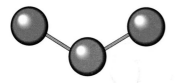

8.7 ACTIVITY FORMATION OF ACETYLENE, A COVALENT COMPOUND

A welder working with an acetylene torch on a pipeline

Covalently bonded compounds make up an incredibly large and important group of compounds. Examples include the oxygen that we breathe, the carbon dioxide that plants use in photosynthesis, the nitrogen that makes up about eighty percent of the air in the atmosphere, and the water that we drink.

Many biologically important compounds are covalently bonded. Common sugar is an example. It is an important energy source for the human body. In addition to sugar there are many other biologically important compounds such as proteins, fats, and carbohydrates. All of these compounds are necessary for the proper maintenance and functioning of the human body.

Another very large and important class of covalently bonded compounds is the hydrocarbons. Hydrocarbons are used as fuels; you will probably recognize some of the more familiar names, such as propane, octane, and butane.

In this activity you will produce a hydrocarbon called acetylene. Acetylene is a fuel used in welding torches. When it combines with oxygen, it produces a very hot flame that can melt certain metals.

Since welding often involves the joining of pieces of pipe, the welder must be able to add a piece of metal that can be melted around the spot where the pipe is to be joined. After the metal is melted by the acetylene torch, the torch is removed, the metal cools, and the two pieces of metal become one solid piece.

When acetylene burns in pure oxygen, it produces carbon dioxide. This is called complete combustion. However, if there is not enough oxygen supplied during combustion, additional products are formed aside from carbon dioxide. This is called incomplete combustion. In this activity you will see both complete and incomplete combustion.

Materials

burner	250 mL beaker
forceps	tap water
wooden splints	calcium carbide lumps
test tubes (4)	limewater

Method

1. Invert a test tube full of water into a beaker of water, making sure that no water runs out of the test tube.

2. Use the forceps to place a few small pieces of calcium carbide into the beaker. Observe the bubbles rising from the calcium carbide. These bubbles are acetylene gas.

3. Place the inverted test tube over the calcium carbide and let the acetylene gas completely fill the test tube.

4. Remove the test tube from the water and place it mouth down on the desk top.

5. Repeat steps 1–4 three more times with the three other test tubes, collecting one-half, one-third, and one-twelfth of a test tube of acetylene, respectively.

6. Lift each test tube from the water, always keeping the test tube inverted. Allow air to replace the water that runs out of the test tubes from step 5.

7. Place your thumb over the end of each test tube. Shake each vigorously for 20–30 s to mix the air and the acetylene. Return each test tube mouth-down onto the desk top.

8. Hold each test tube in a horizontal position and bring a burning splint close to its mouth. Wait until each reaction is complete. Place each test tube mouth-down on the desk top. Observe each test tube carefully and make diagrams of what you observe.

9. Add 2–3 mL of limewater to each test tube. Place your thumb over the end of each test tube and shake vigorously for several seconds. Record your observations.

Questions and Conclusions

1. Which test tube held the least oxygen? the most?

2. What do the results of the limewater test indicate?

3. In which test tube did complete combustion occur? How could you tell?

4. In which test tube did incomplete combustion occur? What other products were formed?

5. Acetylene is covalently bonded. The atoms in acetylene are as follows: **H C C H**. Draw the structural formula for this molecule.

6. Which type of multiple bond exists in acetylene?

7. When welders first ignite an acetylene torch, the flame is often orange-colored and sooty. Why? What needs to be done to get the proper flame for welding?

8.8
NAMING COVALENT COMPOUNDS

Covalent bonds are formed when a nonmetal bonds with another nonmetal. Because the Stock System, which uses Roman numerals, can lead to ambiguity, a different method is used to name covalent molecules. The number of each type of atom in the molecule is indicated by a prefix. The Greek prefixes used are shown in Table 8.4.

Table 8.4   Commonly used numerical prefixes

Number of Atoms	Prefix
1	mon(o)
2	di
3	tri
4	tetr(a)
5	pent(a)
6	hex(a)

A prefix is used in front of each element. The second element in the formula, the more electronegative element, ends in **ide**. For example,

P_2O_3 is called **di**phosphorous **tri**ox**ide**
N_2O_5 is called **di**nitrogen **pent**ox**ide**
CO_2 is called **mono**carbon **di**ox**ide**

If there is only one atom of the first element in the formula, the prefix mon(o) is usually not used but is understood to be in place. This is why CO_2, monocarbon dioxide, is more commonly called carbon dioxide.

The vowel of a prefix is sometimes dropped when it comes before another vowel. For example "monooxide" is written monoxide, and "pentaoxide" is written pentoxide.

Figure 8.5   The nonmetals that commonly form covalent compounds

Try These

1. Name the following covalent compounds:
   a. P_2O_5 c. SO_2 e. CS_2 g. CO
   b. N_2O d. H_2O f. P_2S_3 h. P_2O_3

2. Write the formula for the following covalent molecules:
   a. dinitrogen tetroxide
   b. sulfur trioxide
   c. phosphorus pentachloride
   d. sulfur hexafluoride
   e. carbon tetrachloride
   f. iodine monochloride
   g. diphosphorus trisulfide
   h. dinitrogen trioxide

8.9 ACTIVITY PROPERTIES AND USES OF POLAR AND NONPOLAR MOLECULES (SOLVENTS)

You have learned that covalent bonds are formed by the sharing of electrons. The sharing can be equal or unequal. This depends on the electronegativities of the atoms involved. The bonds may be polar or nonpolar. Molecules may be polar or nonpolar depending on the types of bonds and the shape of the molecule.

In this activity you will carry out a simple test for a liquid polar substance and use these substances to study the solubility of polar substances.

Part A

Materials	
burettes (2)	animal fur
beakers (2)	water
plastic ruler	kerosene
silk cloth	retort stand
ebonite rod	burette clamp

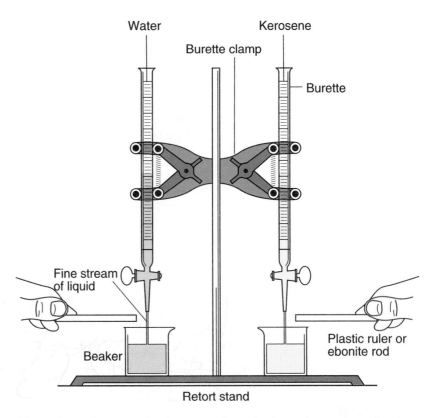

Figure 8.6   Apparatus for the comparison of polar and nonpolar molecules

Method

1. Prepare a chart like the one shown to record your observations.

Test	Water	Kerosene
plastic		
ebonite		

2. Set up the apparatus as shown in Figure 8.6.
3. a. Rub the plastic ruler with silk to produce a charge on the ruler.
   b. Open the burette containing water to allow a fine stream of water to run into the beaker.
   c. Put the plastic ruler near the stream of water. Record your observations.
4. Repeat step 3 using an ebonite rod rubbed with animal fur to produce the charge in step 3a.
5. Repeat steps 3 and 4 but this time using the burette containing kerosene instead of water in 3b.

Questions and Conclusions

1. Which substance is affected more by the charged rods?
2. Which is the polar substance?
3. Explain why the polar substance is affected in a similar way by differently charged rods.

Part B

The polar or nonpolar nature of a molecule has an important influence on its behavior. For example, a common ionic solute like salt will dissolve readily in some solvents but not in others. What factors determine whether a solute will or will not dissolve in a particular solvent?

You may have noticed that some paints are oil-based and some are water-based. Water is used to clean brushes and rollers used with water-based paints. Paint thinner is used for oil-based paints. Both water and paint thinner are solvents. One is polar, the other is nonpolar.

In this activity you will try to dissolve a polar compound, glycerol, and a nonpolar compound, iodine, in different solvents. The solvents you will use for most of the activity are water, which is polar, and kerosene, which is nonpolar.

A solid solute that dissolves in a solvent is said to be soluble. A different term is used when the solute is liquid. A liquid solute that dissolves in a solvent is said to be miscible. Miscible means that the solute is soluble, or completely mixed, in the solvent. A liquid solute that is not soluble in the solvent is said to be immiscible

At the end of this activity you will use an **emulsifying agent** to make an immiscible mixture miscible. The resulting mixture is called an emulsion.

Materials

test tubes (8)	water
test tube rack	glycerol
stoppers (7)	unknown solid
medicine dropper	vegetable oil
kerosene	vinegar
solid iodine	egg yolk

Method

1. Prepare a data table in your notebook.
2. Fill three different test tubes one-third full of kerosene. Label the test tubes.
3. Fill three different test tubes one-third full of water. Label the test tubes.
4. Add one or two small iodine crystals to one test tube of kerosene and to one test tube of water. Stopper the test tubes and shake well.
5. Observe carefully to see whether the solute dissolves in either or both of the solvents.
6. Repeat steps 4 and 5 with glycerol.
7. Repeat steps 4 and 5 with the unknown solid.
8. Add one-third of a test tube of vinegar to one-third of a test tube of vegetable oil. Stopper the test tubes and shake well. Observe whether the two are miscible or immiscible.
9. Add a few drops of egg yolk to the mixture from step 8. Stopper the test tubes and shake well. Observe whether the two are miscible or immiscible.

Questions and Conclusions

1. Iodine molecules are nonpolar. In which solvent was the iodine soluble?
2. Glycerol is polar. In which solvent did glycerol dissolve?
3. Is your unknown solute polar or nonpolar? What evidence do you have?
4. In what type of solvents are polar solutes most soluble?

5. In what type of solvents are nonpolar solutes most soluble?
6. What can you conclude about the solubility of polar or nonpolar solutes in polar or nonpolar solvents?
7. Were the oil and vinegar miscible or immiscible when they were first mixed?
8. Were the oil and vinegar miscible or immiscible after the egg yolk was added? What food is made this way?
9. What is the emulsifying agent in this activity?

8.10 ACTIVITY COMPARING THE PROPERTIES OF IONIC AND COVALENT COMPOUNDS

In Chapter 7 you studied some of the properties of ionic compounds. In this activity you will compare the properties of ionic compounds and covalent compounds.

Materials

burner	sodium chloride (A)
conductivity apparatus	potassium iodide (B)
scoopula	potassium chloride (C)
deflagrating spoons (2)	sugar (D)
watch glass	benzoic acid (E)
beakers (7)	camphor (F)
unknown solid	

Method

1. Prepare a chart similar to the one shown.

Physical Property	Ionic			Covalent			Unknown
	A	B	C	D	E	F	
odor							
melting point							
hardness							
conductor or nonconductor							

2. Try to detect whether any of the substances have an odor. Record your observations.

3. Ignite a burner. Place a small sample of one ionic substance (A) and one covalent substance (D) in two different deflagrating spoons. Hold both samples in front of the burner flame until one of them melts. Record the melting point of the substance that melted as *low* and the melting point of the one that did not as *high*.

4. Repeat step 3 for the second ionic and covalent compounds (B and E).

5. Repeat step 3, using the final pair of compounds (C and F).

6. Determine whether each substance feels hard or soft. Do this by first rubbing a small sample of each between your fingers. Check your observations by trying to crush a few crystals of each. Place each compound in the watch glass one at a time and use a scoopula to try to crush them.

7. Try to make a solution of each compound in water using separate beakers.

8. Use a conductivity apparatus to test the solutions of each of the six substances.

9. Repeat steps 3–8 for the unknown substance.

Questions and Conclusions

1. Which compounds have noticeable odors? What type are they?

2. Which compounds have high melting points? What does this tell you about the strength of ionic bonds?

3. Are ionic compounds harder or softer than covalent compounds? What does this tell you about the strength of ionic bonds?

4. Are solutions of covalent compounds good conductors of electricity?

5. Was your unknown substance an ionic or covalent compound?

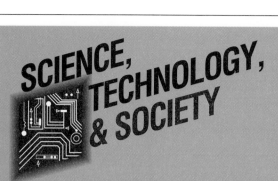

SCIENCE, TECHNOLOGY, & SOCIETY

Chlorofluorocarbons and Ozone

Chlorofluorocarbons

Methane, CH_4, is a covalent compound. Carbon tetrachloride, CCl_4, is also a covalent compound in which chlorine has replaced hydrogen. Other substitutes and combinations of substitutes can be made for hydrogen. Two examples include CCl_3F and CCl_2F_2. These two gaseous compounds are called Freons and belong to a group of covalent chemicals called chlorofluorocarbons (CFCs).

One property of CFCs is that they have a high critical temperature. This makes them useful in refrigeration and air conditioning systems.

Evaporation of a liquid is an endothermic process, that is, it takes in heat. Condensation of a gas, the opposite, releases heat. This process is called exothermic.

In a refrigerator, the compressor heats the Freon gas. The hot compressed gas gradually cools as it circulates through the coils on the back of the refrigerator. (It gives off heat which explains why the coils are always warm). As the gas cools beyond its critical temperature, it liquefies and gives off its heat of vaporization.

The liquid Freon then passes through a valve into a low pressure region inside the refrigerator box. Under low pressure the liquid evaporates. As it evaporates it absorbs its heat of vaporization, cooling the inside of the refrigerator. When the gas reaches the compressor, the cycle is repeated. Heat is continually taken from the inside of the refrigerator and released to the outside.

Freons have some other applications: they have been used in aerosol cans as propellants, in making polystyrene, and as solvents in the electronics industry. They also destroy ozone, a very important molecule in the upper atmosphere.

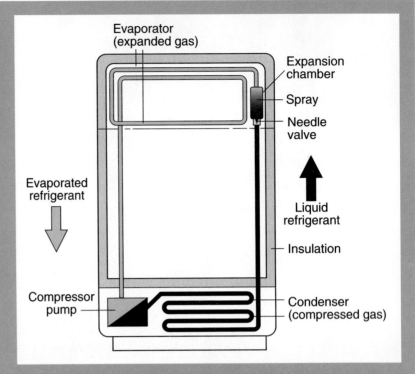

Figure 8.7   How a refrigerator works

Ozone

Over the last few decades, the ozone hole above the Antarctic has grown.

Ozone, O_3, in the atmosphere near the earth's surface is a pollutant. It is a toxic substance and a major contributor to photochemical smog. It can irritate eyes and lungs in concentrations of 0.5 ppm, which are sometimes reached in smog covering large cities like Los Angeles.

However, in the stratosphere, 16–48 km above the earth, it forms a shield protecting the earth from harmful ultraviolet radiation. A drop in the ozone layer concentration could result in increases in skin cancer among humans and damage to other living things.

Through chemical reactions in the atmosphere, ozone is being continually destroyed and replenished. The atmosphere consists of 78% nitrogen and 20% oxygen, mostly in the form of O_2, the molecule we breathe. Ultraviolet light from the sun breaks apart the O_2 molecule and the single atoms produced can combine with other O_2 molecules to produce ozone, O_3. Sometimes ultraviolet light breaks a bond in the ozone molecule. In this case the oxygen atom produced can combine with another ozone molecule to produce two oxygen molecules, eliminating the ozone.

In the 1970s scientists discovered that compounds containing chlorine, nitrogen, or hydrogen could act as catalysts for the destruction of ozone in the atmosphere.

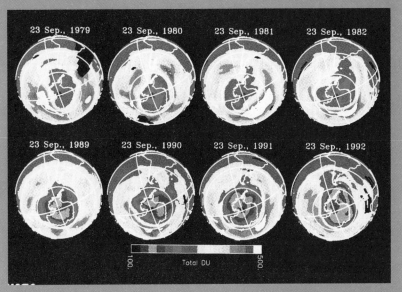

These eight false-color images show the total ozone from Antarctica in two four-year periods (1979–1982 and 1989–1992). The color scale shows total ozone values. Reds and oranges indicate high ozone concentrations, while the blues and purple indicate low ozone concentrations. Black unfilled areas are missing data. The white areas are high ozone values.

In 1974 Molina and Roqland of the University of California showed that a group of industrial chemicals, chlorofluorocarbons (CFCs), could destroy ozone.

CFC molecules take at least ten years to rise through the lower atmosphere to the stratosphere. In the stratosphere, ultraviolet light can free chlorine atoms from CFCs. Chlorine, as a single atom or in combination with oxygen in chlorine monoxide, **ClO,** is an effective agent for destroying ozone. Each chlorine atom can destroy thousands of ozone molecules through a chain reaction.

In 1978 the U.S. government banned certain chlorofluorocarbons from aerosol use. In 1988 E.I. duPont de Nemours and Company began to phase out the production of CFCs.

Dry Cleaner

Dry cleaners must have a knowledge of fabrics, colors, solvents, solutions, and the cleaning process. A dry cleaner separates clothing into categories for different treatments. Solvents are applied to the clothing to remove various stains. The dry cleaner controls the types and amounts of detergents, bleaches, and solvents used. The solvents can usually be reused.

Dry cleaners satisfy their customers by using the proper solvents to remove stains and restore valuable garments that could not be cleaned by traditional methods.

Exercises

Key Words

complete combustion
covalent bond
covalent molecule
diatomic molecule
electronegativity
emulsifying agent
immiscible
incomplete combustion
miscible
multiple bond
nonpolar covalent bond
nonpolar molecule
polar covalent bond
polar molecule
single bond
structural formula

Review

1. Define "covalent bond," "diatomic molecule," and "stable octet."
2. What is the difference between a covalent bond and an ionic bond?
3. Draw electron dot diagrams for the formation of
   a. a hydrogen gas molecule
   b. an oxygen gas molecule
   c. a nitrogen gas molecule
4. Draw Lewis structures and structural formulas and write the formulas for the following molecules. Be careful to include double and triple bonds.

   a. **Br Br**   e. **H O H**   i. **O C O**

   b. **H P H H**   f. **H Cl**   j. **H C N**

   c. **Br Br C Br Br**   g. **H N H H**   k. **N N**

   d. **H S H**   h. **H H C H H**   l. **S C S**

5. Show the electron dot diagram for a hydrogen fluoride molecule. Is this molecule polar or nonpolar?
6. Show that hydrogen chloride is a polar molecule.
7. Is the bonding in lithium fluoride polar or ionic? Why?
8. Draw structural formulas for a molecule of
   a. hydrogen gas   c. hydrogen fluoride
   b. water   d. ammonia (**NH$_3$**)
9. Name the following covalent molecules:
   a. **SiO$_2$**   b. **NO**   c. **BF$_3$**   d. **ICl$_3$**   e. **H$_2$S**   f. **SO$_2$**
10. Write the formula for the following covalent molecules:
    a. phosphorus trichloride   c. sulfur tetrafluoride
    b. dichlorine monoxide   d. dinitrogen trioxide
11. a. Grease, a nonpolar substance, stains your clothing. Would a polar or nonpolar solvent be more efficient in removing the stain? Why?
    b. In what way would an emulsifying agent be useful?

Extension

1. Build large-scale models of some covalent compounds using foamed polystyrene spheres and toothpicks. Paint them and use them to decorate your classroom.
2. Investigate the different types of solvents used to remove stains in dry cleaning. Classify them as polar or nonpolar.
3. Silk screening uses the idea that "like dissolves like." Prepare a demonstration or activity for the class.

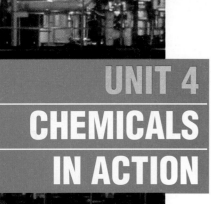

CHAPTER 9
CHEMICAL EQUATIONS AND REACTIONS

Chemists find that the information in a balanced chemical equation is very useful. A chemical equation is a short way to represent a chemical reaction. It is a form of chemical "shorthand."

Chemists use these equations to predict what will be produced and how much will be produced. In addition, the amounts of reactants needed and the conditions needed to carry out the reactions are of great interest. In industry, this information is important to decide whether a reaction will be commercially profitable.

The types of chemical equations studied in this chapter are the basis of most of the applications of chemistry found throughout this text.

Key Ideas

- Chemical reactions can be identified by the changes that substances undergo during a chemical reaction.

- Chemical equations are the language of chemistry.

- All chemical equations follow the Law of Conservation of Mass.

- A balanced chemical equation contains a great deal of useful information.

- The mole is the unit of amount of substance used by chemists.

9.1
CHEMICALS
IN ACTION

As lead nitrate is poured into potassium iodide, a yellow precipitate, lead iodide, is formed.

When a chemical reaction takes place, the properties of the product(s) are different from those of the reactant(s). This type of change is called a chemical change.

It is usually, but not always, obvious that a chemical change has taken place. One of the following is usually observed during a reaction if a chemical change is taking place:

1. increase or decrease in temperature
2. release of a gas
3. formation of a precipitate
4. change in color
5. production of an odor or change in an odor

Try These

1. Which of the following involve chemical changes? What evidence supports your choices?
   a. the burning of gasoline
   b. the freezing of water
   c. magnesium burning in oxygen to form magnesium oxide
   d. methane melting at -182°C
   e. boiling an egg
   f. liquid solutions of potassium chloride plus silver nitrate combining to form solid silver chloride
   g. nitroglycerin exploding on impact or when heated
   h. water boiling at 100°C

2. Not all of the items in question 1 are examples of a chemical change. What kind of change are they?

9.2 ACTIVITY
WHAT HAPPENED
TO THE VINEGAR?

Many substances found in the kitchen react with each other chemically. In this activity you will look for evidence that tells you that a chemical reaction is occurring.

Materials

glass stirring rod
100 mL beakers (2)
test tubes (2)
measuring spoons

baking soda
cream of tartar
vinegar

Reactions with household products

Method

1. Add 10 mL of vinegar to each of two 100 mL beakers. Note the odor.
2. Add 4 mL of cream of tartar to the first beaker. Stir well and record your observations.
3. Add 4 mL of baking soda to the second beaker. Stir well and record your observations.
4. Decant the liquid from each beaker into labeled test tubes. Note the odor of each.
5. Add a pinch of baking soda to each test tube and record your observations.

Questions and Conclusions

1. In each case, did a chemical reaction occur?
2. If a chemical reaction occurred, what evidence indicated this?
3. If a chemical reaction did not occur, what test indicated this?

9.3 ACTIVITY
CONSERVATION
OF MASS

In the 1780s, Antoine Lavoisier, a French chemist, performed some quantitative experiments involving mercury and air. The product of the reaction was a red compound. He found that the increase in mass in reacting mercury with air to produce the red product was equal to the amount of oxygen produced when the reaction was reversed. Lavoisier's experiments demonstrated the Law of Conservation of Mass. In this activity you will observe a chemical reaction and test to see whether mass is conserved.

Materials

balance	small test tube
250 mL Erlenmeyer flask	sodium sulfate solution
stopper for flask	barium nitrate solution

Stopper

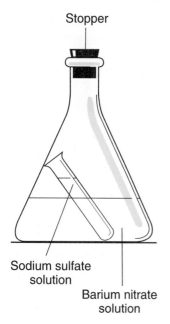

Sodium sulfate
solution

Barium nitrate
solution

Method

1. Put about 20 mL of barium nitrate solution in the Erlenmeyer flask.
2. Fill the small test tube with sodium sulfate solution. Make sure that the outside of the test tube remains clean and dry.
3. Carefully put the small test tube inside the flask as shown.
4. Carefully stopper the flask.
5. Measure the mass of the stoppered flask with the test tube and solutions.
6. Gently invert the flask and allow the chemicals to mix. Record your observations.
7. Measure the mass of the stoppered flask with the test tube and products again.

Questions and Conclusions

1. Did a chemical reaction occur?
2. If a chemical reaction occurred, what evidence indicated this?
3. If a chemical reaction did not occur, what evidence indicated this?
4. What information about conservation of mass did this activity provide?

9.4
WRITING BALANCED CHEMICAL EQUATIONS

A chemical equation is a description of a chemical reaction. It is written in chemical symbols instead of words.

Every chemical equation is written in the same way:

1. Reactants are written first (before the arrow).

2. Products are written second (after the arrow).

3. The arrow separating reactants and products means "yields" or "produces."

4. Individual reactants and products are separated by "plus" signs.

5. The symbols (g), (s), and (l) are used to represent the physical state of the substance. Aqueous (aq) means dissolved in water.

The word equation for the formation of hydrogen chloride gas is

hydrogen gas plus **chlorine gas** produces **hydrogen chloride gas**

The following is the chemical equation for the same reaction:

$$H_2(g) + Cl_2(g) \longrightarrow HCl(g) \text{ (unbalanced)}$$

This equation says that one molecule of hydrogen gas combines with one molecule of chlorine gas to produce one molecule of hydrogen chloride gas.

Activity 9.3 showed that mass is neither gained or lost during a chemical reaction. Chemical reactions follow the Law of Conservation of Mass. This means that the number of each type of atom in the reactants must equal the number of atoms of each type in the products.

The unbalanced equation for the chemical reaction above does not follow the Law of Conservation of Mass. It shows that there are two atoms of hydrogen and two atoms of chlorine on the reactant side but only one atom of hydrogen and one atom of chlorine on the product side. This equation is unbalanced and is called a skeleton equation

To balance the skeleton equation, there must also be two atoms of hydrogen and two atoms of chlorine on the product side of the equation. This means that there must be two molecules of hydrogen chloride product. The number of molecules is represented by the coefficient, the number in front of the formula for the substance. The coefficient of the hydrogen chloride molecule is changed from 1 to 2:

$$1\,H_2(g) + 1\,Cl_2(g) \longrightarrow 2\,HCl(g) \text{ (balanced)}$$

Note that the coefficient is not usually written in front of the element or molecule if the coefficient is 1, so that the equation is actually written

$$H_2(g) + Cl_2(g) \longrightarrow 2\,HCl(g)$$

This is now a balanced chemical equation. There are two hydrogen atoms and two chlorine atoms on each side of the equation.

The subscripts following the atoms cannot be changed to balance a chemical equation. These numbers give the number of atoms of each type present in one molecule of a substance. Changing the subscript would alter the molecular structure of the substance. For example, O_2 is a molecule of oxygen, but O_3 is a molecule of ozone; H_2O is a molecule of water, but H_2O_2 is a molecule of hydrogen peroxide.

A chemical equation can be balanced only by changing the coefficient or number of molecules present.

In summary, the following are the steps in writing balanced chemical equations:

Step 1

Write the chemical formula for each reactant and product followed by the state of each: solid (s); liquid (l); gas (g); aqueous (dissolved in water) (aq).

Step 2

Adjust the number of molecules until there are the same number of atoms of each type on both sides of the equation. This balances the mass of both reactants and products.

Do not change the subscripts in a formula to balance an equation. Changing these numbers changes the structure of the molecule.

Try These

1. What information must you know to write a chemical equation?
2. On which side of the equation are the products written?
3. Why must chemical equations follow the Law of Conservation of Mass?
4. What do the symbols (g), (l), (s), and (aq) mean?
5. What is a skeleton equation?
6. How is a chemical equation balanced?
7. Can the subscripts in a chemical formula be changed in order to balance a chemical equation? Why or why not?
8. Balance the following equations:
    a. $H_2(g) + Br_2(g) \longrightarrow HBr(g)$
    b. $Na(s) + Cl_2(g) \longrightarrow NaCl(s)$
    c. $Na(s) + O_2(g) \longrightarrow Na_2O(s)$

d. $Zn(s) + HCl(aq) \longrightarrow ZnCl_2(aq) + H_2(g)$

e. $Fe(s) + O_2(g) \longrightarrow Fe_2O_3(s)$

9. Write word equations for the chemical equations presented in question 8.

10. Balance the following equations:

a. $P_4(s) + O_2(g) \longrightarrow P_2O_5(s)$

b. $Al_2O_3(s) + H_2(g) \longrightarrow Al(s) + H_2O(l)$

c. $Ca(s) + O_2(g) \longrightarrow CaO(s)$

d. $Cl_2(g) + AlBr_3(aq) \longrightarrow Br_2(l) + AlCl_3(aq)$

e. $HgO(s) \longrightarrow Hg(l) + O_2(g)$

f. $Fe(s) + H_2O(l) \longrightarrow Fe_2O_3(s) + H_2(g)$

11. Write word equations for the chemical reactions presented in question 10.

12. Write balanced chemical equations for the following chemical reactions:

a. aluminum metal plus hydrogen chloride gas yields solid aluminum chloride plus hydrogen gas

b. zinc metal plus oxygen gas produces solid zinc oxide

c. magnesium metal plus gaseous carbon dioxide yields solid magnesium oxide plus solid carbon

d. When zinc metal and sulfur powder are heated, they form solid zinc sulfide

9.5 ACTIVITY BALANCING EQUATIONS WITH GUMDROPS

In Section 9.4, you learned that chemical equations provide a great deal of information. You have also learned that chemical reactions follow the Law of Conservation of Mass, so that equations representing reactions must be balanced.

In this activity different-colored gumdrops will be used to illustrate

a. the rearrangement of atoms in a chemical equation

b. that balanced equations follow the Law of Conservation of Mass

Materials

toothpicks

gumdrops (5 colors)

Method

1. Make models to illustrate the following reactions. Connect the "atoms" using toothpicks as "bonds." Be sure to balance the equations by having the same number of each type of "atom" on each side of the equation. The first one has been done for you as an example.

   a. $H_2 + O_2 \longrightarrow H_2O$

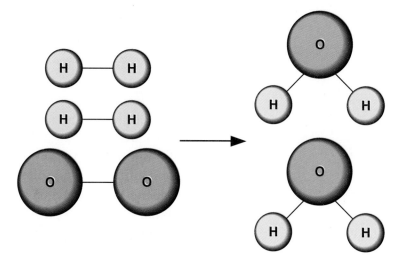

   b. $Mg + O_2 \longrightarrow MgO$
   c. $H_2CO_3 \longrightarrow H_2O + CO_2$
   d. $KI + Br_2 \longrightarrow KBr + I_2$
   e. $Zn + HCl \longrightarrow ZnCl_2 + H_2$
   f. $FeS + HCl \longrightarrow FeCl_2 + H_2S$

9.6

FORMULAS, THE MOLE, AND MOLAR MASS

Chemists often need to measure out a certain mass of a substance. To do this they use a unit of amount of substance called the **mole** (symbol "mol"). The mole concept came about as a result of the work of an Italian physicist, Amadeo Avogadro. From his experiments with gases in the early 1800s, he developed the idea of a mole as a certain number of particles of any substance. That number of particles of a substance has a certain mass that chemists use to measure a mole of that substance. Whenever atoms or molecules combine with one another, they combine in simple whole number, or mole, ratios.

For monatomic elements, one mole has the same mass as the atomic mass of the element, but it is measured in grams instead of atomic mass units. The mass of one mole of a substance is called its molar mass

The following chart presents a few examples of the atomic masses of monatomic elements and their corresponding molar masses.

Element	Atomic Mass	Molar Mass
H	1 u	1 g
C	12 u	12 g
O	16 u	16 g
Ar	40 u	40 g

One mole of various substances: (top) sucrose (342.30 g),
(middle, left to right) lead (207.21 g), potassium dichromate (294.21 g),
mercury (200.61 g), water (18.016 g), (bottom, left to right) copper (63.54 g),
salt (58.45 g), sulfur (32.066 g).

As you have learned, gaseous elements are not found as atoms in nature but most often as diatomic molecules. The following are a few examples of molecular and molar masses for this type of molecule.

Molecule	Molecular Mass	Molar Mass
H_2	2(1 u) = 2 u	2(1 g) = 2 g
F_2	2(19 u) = 38 u	2(19 g) = 38 g
Cl_2	2(35.5 u) = 71 u	2(35.5 g) = 71 g

Binary compounds are also very common. These occur when two different elements combine. To calculate the formula mass and molar mass of this type of compound, the number of atoms present is also multiplied by the mass of each element, in atomic mass units or grams. Then the mass of the different elements is added together. The following are a few examples.

Substance	Formula Mass	Molar Mass
NaCl	1(23 u) + 1(35.5 u) = 58.5 u	1(23 g) + 1(35.5 g) = 58.5 g
H_2O	2(1 u) + 1(16 u) = 18 u	2(1 g) + 1(16 g) = 18 g
Al_2S_3	2(27 u) + 3(32 u) = 150 u	2(27 g) + 3(32 g) = 150 g

Try These

1. Name the units in which molecular mass is measured.
2. Name the units in which molar mass is measured.
3. What does molar mass mean?
4. State the atomic mass for each of the following:
   a. Na   c. Be   e. Ca   g. Mg
   b. K   d. B   f. Al   h. Ne
5. What is the formula mass or molecular mass for each of the following?
   a. CO   d. MgO   g. K_2O j. CH_4
   b. CO_2 e. CaO   h. NaF   k. KCl
   c. CCl_4 f. $MgCl_2$ i. SO_2 l. SiF_4
6. What is the mass of one mole for each item in question 5?
7. How many atoms are present in each of the molecules in question 5?
8. What information do coefficients in a chemical reaction give us?

9.7 ACTIVITY

HOW MANY MOLES?

Sometimes chemists also want to know how many moles of a substance are present in a given mass of the substance. This is easy to calculate. For example, if you wanted to find out how many moles of water were present in a 90 g sample of water, you would calculate as follows:

Let "n" equal the number of moles present.
n = number of moles
mass of the substance (water) = 90 g
molar mass of water = 18 g

$$n = \frac{\text{mass of the substance}}{\text{molar mass of the substance}}$$

$$= \frac{90 \text{ g}}{18 \text{ g/mol}}$$

$$= 5 \text{ mol}$$

There are 5 mol present in a 90 g sample of water.

Materials

150 mL beakers (2)
sodium chloride
aluminum strip

Method

1. Measure the mass of an empty beaker.
2. Fill the beaker with approximately 100 mL of water.
3. Find the mass of the water and beaker and calculate the mass of the water.
4. Calculate how many moles of water are in the beaker.
5. Repeat steps 1–4 using sodium chloride instead of water.
6. Find the mass of the aluminum strip.
7. Calculate the number of moles of aluminum present.

Questions and Conclusions

1. Calculate the number of moles in
   a. 10 g of $H_2(g)$
   b. 100 g of $O_2(g)$
   c. 10 g of $NaCl(s)$
   d. 10 g of Al_2S_3

2. Copy the following chart into your notebook and complete it by filling in the blanks. **Do not write in this textbook.**

Substance	Formula	Molar Mass	Mass	Number of Moles
sodium chloride				2.0
hydrogen gas			20.0 g	
aluminum			2.7 g	
magnesium oxide				2.5

9.8

THE INFORMATION IN A CHEMICAL EQUATION

Chemists use the information in a balanced chemical equation to determine what mass of each substance is needed for a particular reaction to occur.

In Sections 9.4–9.7, you learned that substances combine in simple mole ratios. The number of moles of each reactant needed and their molar masses provide us with important information. For example, the chemical reaction for the formation of water gives us the following information:

$$2\ H_2(g) + O_2(g) \longrightarrow 2\ H_2O(g)$$
$$2\ \text{molecules}\ H_2 + 1\ \text{molecule}\ O_2 \longrightarrow 2\ \text{molecules}\ H_2O$$
$$2\ \text{mol}\ H_2 + 1\ \text{mol}\ O_2 \longrightarrow 2\ \text{mol}\ H_2O$$
$$(2\ \text{mol})(\text{molar mass}\ H_2) + (1\ \text{mol})(\text{molar mass}\ O_2) \longrightarrow (2\ \text{mol})(\text{molar mass}\ H_2O)$$
$$(2)(2\ g) + (1)(32\ g) \longrightarrow (2)(18\ g)$$
$$4\ g\ H_2 + 32\ g\ O_2 \longrightarrow 36\ g\ H_2O$$

The coefficient in front of the molecule indicates how many molecules, and therefore how many moles, are present. The Law of Conservation of Mass is always obeyed. It provides a way of checking your work. The mass of the reactants equals the mass of the products. In this case, the mass of the reactants is 36 g (4 g + 32 g) and the mass of the product is 36 g.

Another example is the formation of magnesium oxide. The balanced equation is

$$2\ Mg(s) + O_2(g) \longrightarrow 2\ MgO(s)$$

Therefore:

$$2\ \text{mol}\ Mg + 1\ \text{mol}\ O_2 \longrightarrow 2\ \text{mol}\ MgO$$
$$(2)(24\ g) + (1)(32\ g) \longrightarrow (2)\ (40\ g)$$
$$48\ g\ Mg + 32\ g\ O_2 \longrightarrow 80\ g\ MgO$$

In summary, a balanced chemical equation contains a great deal of useful information about the reactants and products of a chemical reaction. The following information can be obtained from a balanced chemical equation:

1. The numbers and types of substances needed to cause a particular reaction.
2. The numbers and types of substances produced in a particular reaction.
3. The number of moles of each reactant needed.
4. The number of moles of each product produced.
5. The molar mass of each reactant needed.
6. The molar mass of each product formed.
7. The physical state of each of the reactants and products.

Try These

1. For the following equations, state the number of molecules and the number of moles present for each reactant and product.
   a. $H_2(g) + Br_2(g) \longrightarrow 2\ HBr(g)$
   b. $2\ Ca(s) + O_2(g) \longrightarrow 2\ CaO(s)$
   c. $Zn(s) + 2\ HCl(aq) \longrightarrow ZnCl(aq) + H_2(g)$
   d. $4\ Fe(s) + 3\ O_2(g) \longrightarrow 2\ Fe_2O_3(s)$
2. Show that the Law of Conservation of Mass is obeyed for each equation in question 1.

9.9 ACTIVITY
TESTING
FOR DIFFERENT
CHARGES

Sometimes the same substance can take part in more than one reaction with another substance. Examples of this are the two oxides of carbon. One is carbon monoxide, which is a component of car exhaust fumes and is a deadly poison to human beings. The other is carbon dioxide, which we exhale with every breath.

The two oxides of carbon, carbon monoxide (left) and carbon dioxide (right), have different chemical properties.

In this activity you will test for differently charged ions of the same element. Because of the different charges, the chemical reactions and the compounds formed will be different.

Materials

test tubes (2)	iron(III) chloride
stoppers (2)	sodium hydroxide solution
medicine droppers (2)	iron(II) sulfate
distilled water	3% hydrogen peroxide solution

Caution: Sodium hydroxide is corrosive and causes burns. Handle it with care. Wear goggles and an apron. If you spill any on yourself, rinse it off immediately with large amounts of cold water and have your partner tell your teacher.

Method

1. Place a small amount of iron(III) chloride into 2 mL of distilled water in a test tube.
2. Add a few drops of the sodium hydroxide solution. Stopper the test tube and shake. Record your observations.
3. Repeat steps 1–2 using iron(II) sulfate.
4. Add a small amount of 3% hydrogen peroxide solution to the test tube containing iron(II) sulfate from step 3.
5. Compare the results from step 4 with the results from step 2.

Questions and Conclusions

1. What evidence did you observe that showed the difference between iron(II) and iron(III)?
2. Write a word equation for the following reaction. Iron(III) chloride and sodium hydroxide are the reactants. Iron(III) hydroxide and sodium chloride are the products.
3. a. What change occurred in the charge of iron(II) when hydrogen peroxide was added?
   b. What evidence helped you to come to this conclusion?
4. Write balanced chemical equations for the reactions of the transition metals in each of the following:
   a. Tin reacts with chloride ions to form tin(II) chloride and tin(IV) chloride (2 equations).
   b. Mercury reacts with oxygen gas to form mercury(I) oxide and mercury(II) oxide (2 equations).
   c. Iron reacts with oxygen gas to form iron(II) oxide and iron(III) oxide (2 equations).

SCIENCE, TECHNOLOGY, & SOCIETY

Smog

The word **smog** is used to describe air pollution resulting from a combination of **smoke** and **fog**. This combination occurs in cities bordering water and having high humidity. A serious smog condition developed in London, England in December, 1952. Because of the cold and damp, people continually heated their homes, sending smoke from coal furnaces into the air to mix with the thick fog. Over 4000 premature deaths were attributed to the effects of breathing smog containing sulfur dioxide from sulfur in the coal. Cases of pneumonia and bronchitis increased dramatically.

Photochemical smog is formed when light from the sun starts chemical reactions that produce large amounts of nitrogen oxides, ozone, organic peroxides, and hydrocarbons.

In Los Angeles, California, the geography and climate of the area, and the large number of cars, all contribute to producing photochemical smog. Because the city sits in a depression surrounded by mountains, it often encounters a thermal inversion. Normally, warm air near the earth rises and cold dense air higher in the atmosphere moves downward to replace it. In an inversion, the upper air is warmer and less dense and the air near the earth cannot move up. In Los Angeles it cannot move sideways either because of the mountains. The trapped air becomes polluted with exhaust fumes and the products of photochemical reactions.

Car engines produce nitrogen monoxide that reacts with oxygen in the air to form nitrogen dioxide. Nitrogen dioxide is reddish brown and accounts for the color of photochemical smog. Ultraviolet light breaks bonds in nitrogen dioxide, releasing oxygen atoms that combine with oxygen molecules to produce ozone, a toxic and corrosive substance.

CHAPTER 10

PATTERNS

IN CHEMICAL

REACTIONS

This chapter will introduce you to two of the four basic patterns that most chemical reactions follow. In this chapter you will examine direct combination or synthesis reactions and also their reverse, decomposition.

These reactions, together with the types you will study in Chapter 11, form the basis of most of the applications found throughout this textbook.

Burning and corrosion are two common examples of direct combination reactions. Both combination and decomposition reactions are widely used to produce important industrial chemicals.

Key Ideas

- The pattern of a reaction often allows the prediction of products.
- Grouping reactions into patterns is another way of organizing chemical information.
- Each type of chemical reaction has an important application in industry.
- The types of industries using these reactions provide many different types of jobs.
- These types of chemical reactions play a role in our everyday lives.

Dynamite can be used to destroy unwanted buildings.

10.1
THE HABER PROCESS— A DIRECT COMBINATION REACTION

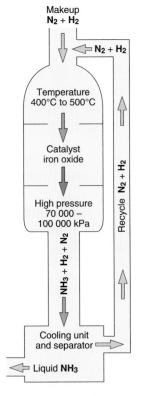

Figure 10.1   A model of the Haber process

A **direct combination** reaction fits the general pattern of

$$A + B \longrightarrow AB$$

The product **AB** is a direct combination or addition of the reactants. Ammonia is made industrially by a direct combination reaction known as the **Haber process**. This process was developed in 1913 by the German chemist Fritz Haber. In this process, nitrogen gas and hydrogen gas are combined at high temperatures and pressures in the presence of a catalyst. A **catalyst** is a substance that can be used to speed up chemical reactions.

The chemical reaction for the production of ammonia is

$$N_2(g) + 3\,H_2(g) \longrightarrow 2\,NH_3(g)$$

Ammonia is a very important chemical in today's society. Billions of kilograms are produced around the world every year. Most of it is used in the production of fertilizer, but it is also used to produce nitric acid, nylon, plastics, livestock feed, smelling salts, and household cleansers.

Figure 10.2   Some products made from ammonia

During World War I, the Haber process was used to make explosives. Two of the major explosives used in that war were nitroglycerin and dynamite. Nitric acid is needed to manufacture these explosives, but Germany's supply of nitrates from Chile had been cut off by the British. Instead of using nitrates, the Germans had to use ammonia to make nitric acid.

Because of the high temperatures and pressures needed, it is not possible to make ammonia in the laboratory. However, many other direct combination or **synthesis** reactions are possible. You will perform some of these in the following activity.

Try These

1. What is the general pattern of a direct combination reaction?
2. What is another name for a direct combination reaction?
3. List four uses of ammonia.
4. What is the main use of ammonia in today's society?
5. Why is the item named in your answer to question 4 so important to the world today? Use the library resource center to research the use of this item in developing countries.

10.2 ACTIVITY
DIRECT
COMBINATIONS

In this activity you will carry out four direct combination reactions.

Materials

burner
tongs
oxygen generator
deflagrating spoon
gas bottles (2)
glass plates (2)
pen
copper wire
copper or silver coin
pH paper or indicator solution
sandpaper
distilled water
magnesium ribbon
cobalt blue glass
sulfur

Caution: Do not look directly at the magnesium while it burns. View the reaction through cobalt blue glass.
Perform steps 5–10 and 19 in a fume hood. Do not breathe the sulfur fumes.

Burning sulfur. This procedure should be done in a fume hood.

Method

1. Prepare a data table to record your observations.
2. Use a pair of tongs to hold a 6-cm piece of copper wire in a burner flame.
3. Take the wire out of the flame and observe any changes.
4. Fill two gas bottles with oxygen as directed by your teacher. Cover both bottles with glass plates.
5. Place a small piece of sulfur on a deflagrating spoon and ignite it in a burner flame.
6. Quickly place the deflagrating spoon in a bottle of oxygen. Cover the bottle with a glass plate and observe the reaction.
7. Cautiously note the odor.
8. After the reaction is complete, remove the deflagrating spoon from the bottle and add distilled water to a depth of 2 cm.
9. Cover the mouth of the bottle with your hand and shake.
10. Remove your hand and test the solution with pH paper or indicator solution.
11. Obtain a 10-cm piece of magnesium ribbon from your teacher.
12. Clean the ribbon by running it through a folded piece of sandpaper.
13. Coil the magnesium ribbon by winding it around a pen, leaving a 1-cm length at the end.
14. Hold one end of the ribbon with tongs and ignite the other end in a burner flame.
15. Quickly put the ribbon into the second bottle of oxygen without touching the sides of the bottle. Do not look directly at the flame. View the reaction through cobalt blue glass.
16. After the reaction is complete, pour distilled water into the bottle to a depth of 2 cm and shake.
17. Test the solution with pH paper or indicator solution.
18. Place a sample of sulfur the size of a pinhead on a copper or silver coin.
19. Using tongs, gently heat the coin until the sulfur melts. Describe the change in appearance of the coin.

Questions and Conclusions

1. Copper(I) oxide is red and copper(II) oxide is black. Which form of copper oxide was produced when the copper wire was heated?
2. Write the balanced chemical equation for the reaction between copper and oxygen to form copper(II) oxide.

3. Why might some of the sulfur not have burned in step 6?

4. What happened when you removed your hand from the mouth of the bottle in which the sulfur was burned? Why?

5. Write both the word and balanced equations for the reaction between sulfur and oxygen to form sulfur dioxide.

6. What effect did sulfur dioxide have on the pH of distilled water?

7. What is the product of the direct combination of magnesium and oxygen?

8. Does the reaction of magnesium and oxygen result in an overall release or absorption of energy?

9. Write both the word and balanced chemical equations for the reaction between magnesium and oxygen.

10. What effect did magnesium oxide have on the pH of distilled water?

11. What property of magnesium makes it useful for camera flashbulbs?

This flashbulb uses magnesium as a light source.

12. What color is the sulfide of silver or copper?

13. Write both the word and balanced chemical equations for the reaction between silver and sulfur.

14. What chemical reaction is taking place when silver tarnishes?

10.3
SULFURIC ACID— A VERY USEFUL CHEMICAL

Almost everyone has experienced the benefits of sulfuric acid. This is the very strong acid found in car batteries. It is one of the most commonly produced industrial chemicals. Each year over 90 million tonnes of sulfuric acid are manufactured worldwide. Beside being used directly in car batteries, it is also used in many processes ranging from the manufacture of plastics to the production of paints.

The first step in the production of sulfuric acid is the same as the synthesis reaction studied in Activity 10.2.

$$S(s) + O_2(g) \longrightarrow SO_2(g)$$

The sulfur for the reaction comes from areas such as Texas where large deposits of natural sulfur are found. In Canada sulfur is obtained mainly from hydrocarbons and metallic sulfides.

Large deposits of elemental sulfur are used as raw material to make chemicals. The building in the background is a chemical plant.

Another synthesis reaction is used in the next step of the process. This reaction involves combining sulfur dioxide with oxygen to form sulfur trioxide.

$$2\,SO_2(g) + O_2(g) \longrightarrow 2\,SO_3(g)$$

A third synthesis reaction completes the production of sulfuric acid. In this reaction, the sulfur trioxide reacts with water to produce sulfuric acid.

$$SO_3(g) + H_2O(l) \longrightarrow H_2SO_4(l)$$

Industrially, sulfuric acid is made by using a process called the contact process. In this process, a catalyst is used to speed up the reaction between sulfur dioxide and oxygen. The final reaction is completed by dissolving sulfur trioxide in a mixture of 98% sulfuric acid and 2% water. Under these conditions the sulfur trioxide reacts with the water to form additional sulfuric acid. If the reaction is attempted by simply adding sulfur trioxide to water, a great deal of heat is generated, making the process commercially impractical.

Sulfur dioxide is released into the atmosphere when fuels containing sulfur impurities are burned, when metal refining processes use ores with sulfur impurities, and also when volcanoes erupt. When this sulfur dioxide combines with water in the atmosphere, one type of acid rain is formed.

Try These

1. What is a common use for sulfuric acid?

2. a. Write the three equations involved in the production of sulfuric acid.

 b. Identify the type of reaction in each case.

 c. Write the general equation for the type of reaction identified in 2b.

10.4

COMBUSTION

AS A DIRECT

COMBINATION

Coal-fired power stations are a major source of electricity.

Three of the reactions you studied in Activity 10.2 were direct combinations between an element and oxygen. These kinds of reactions are called combustion reactions.

Combustion reactions using fuels produce heat energy along with chemical products. In these reactions, the heat energy given off is of most interest to people. The chemical products, which often include water vapor and carbon dioxide, are not usually given much thought.

Coal, a form of carbon, was used during the 1800s and early 1900s as the main source of energy to heat homes. The combustion of coal is an example of a direct combination, as follows:

$$C(s) + O_2(g) \longrightarrow CO_2(g) + \text{heat energy}$$

Coal is still widely used to fuel power stations. The heat energy produced from the direct combination is used to form steam, which drives the turbines in the power plant and produces electricity.

Today, many homes are heated by oil or natural gas. The combustion of each of these fuels is an example of a direct combination reaction with oxygen. In these cases, unlike coal, a compound instead of an element combines with oxygen. However, the end products of all of these reactions are similar. They produce heat energy, carbon dioxide, and water vapor. The combustion reaction for natural gas is

$$CH_4(g) + 2\,O_2(g) \longrightarrow CO_2(g) + 2\,H_2O(g) + \text{heat energy}$$

The general equation in the above case is

$$BC + A \longrightarrow BA + CA$$

Try These

1. List three direct combination reactions that have practical applications.

2. What name is given to combination reactions involving a fuel and oxygen?

3. What is a useful byproduct of a combustion reaction?

10.5 ACTIVITY COMBUSTION, HEAT, AND YOUR ENERGY BILL

Over much of North America we must heat our buildings in the winter to stay warm. Heat is expensive, and producing it usually creates pollution, so wasting energy is something worth avoiding.

In this activity you will learn how to read a gas bill and discuss ways in which energy costs can be reduced.

Method

1. Inspect the gas bill shown in Figure 10.3 and answer the following questions.

GAS BILL

Service to: **John Smith**
55 Our Town

Account Number
41/6255/0123

Description	Amount
Amount previously billed	50.25
Balance forward (Sub-Total)	50.25
Gas Charges - Estimate: Meter Reading not scheduled - (See EBP Below)	
Revised Monthly EBP Amount	72.20
Rental Water Heater	7.30

A REMINDER

The "BALANCE FORWARD" is an unpaid amount from your previous bill. An early remittance will be appreciated. If paid recently, please accept our thanks.

The new Equal Billing Plan (EBP) begins this month. As described in our December billing insert your EBP installments are now calculated over 11 months instead of 12.

Rate	Billing Period Days	Reading Dates		Readings		Consumption	Total Billed
		Previous	Current	Previous	Current		129.75
1A	46	01 JAN 94	15 FEB 94	1035	1046	11 m³	

	Equal Billing Plan				Due Date
Cost of Gas Used This Billing	Cost of Gas Used to Date	Installments Billed to Date	Balance After This Bill is Paid	Amount Paid	01 MAR 94
6.57	6.57	72.20	41.51 CR		Payable After Due Date 132.18

Figure 10.3   A gas bill

Questions and Conclusions

1. What is the billing period represented by this bill?

2. What is the total amount billed for this billing period?

3. How much gas (in m³) was consumed during this billing period?

4. What is the cost of gas consumed during this billing period?

5. What is the advantage of an equal billing plan (EBP)?

6. Name items in a home that might use natural gas.

7. State at least one way of reducing costs and gas consumption for each item listed in question 6.

8. Answer questions 6 and 7 in relation to your school.

9. Research alternative methods of energy production that could reduce energy costs.

A gas meter being read

10.6 ACTIVITY
DECOMPOSITION

Decomposition reactions are the opposite of direct combination reactions. They are usually of the following general type:

$$AB \longrightarrow A + B$$

In this type of decomposition, a substance breaks down into simpler substances or into its elements. Chemists sometimes use heat to cause a substance to decompose. Another way is through spontaneous (or naturally occurring) decomposition.

Hydrogen peroxide (H_2O_2) decomposes spontaneously. In its pure state, it is a colorless, syrupy liquid, 1.5 times as dense as water. Hydrogen peroxide is used for bleaching wood pulp, textiles, and hair, as an antiseptic, and even in rocket fuels. The hydrogen peroxide that we buy in the drugstore is not in its pure state, but is a 3% solution of hydrogen peroxide in water.

Hydrogen peroxide decomposes easily. If left unused for several months, its chemical composition changes, and it can no longer be used for its original purpose. A catalyst may be used to speed up the decomposition.

In this activity hydrogen peroxide and other chemicals will be decomposed. Each follows the general pattern for a decomposition reaction.

Materials

burner or hot plate
evaporating dish
test tube
glass stirring rod
3% hydrogen peroxide solution
carbonated beverage
malachite or copper(II) carbonate
manganese dioxide

Method

1. Prepare a data table to record your observations.
2. Pour 10 mL of 3% hydrogen peroxide solution into a test tube. Add a pinch of manganese dioxide. Note any evidence that a chemical reaction is occurring. Perform a test to determine what gas was released.
3. Stir the carbonated beverage. What happens?
4. Obtain a sample of copper(II) carbonate or malachite, a mineral that contains copper(II) carbonate.
5. Place the sample in an evaporating dish and heat it strongly as you stir it with a glass rod. Record your observations.

Questions and Conclusions

1. What is the general pattern for a decomposition reaction?
2. What liquid do you suspect is left in the test tube after the hydrogen peroxide has decomposed?
3. What gas is released when hydrogen peroxide decomposes?
4. Write the word equation and the balanced chemical equation for the decomposition of hydrogen peroxide.
5. Carbonic acid (H_2CO_3) can be used to make soft drinks bubble and fizz. What gas do you suspect is given off in step 3?
6. Write the word equation and the balanced chemical equation for the decomposition of carbonic acid.
7. When copper(II) carbonate is decomposed, is heat absorbed or released?

Malachite (top) and verdigris (bottom)

8. When malachite or copper(II) carbonate **(CuCO₃)** is heated, a gas is released. What gas do you predict is given off?

9. What is the solid product remaining in the evaporating dish when malachite decomposes?

10. Write the word equation and the balanced chemical equation for the decomposition of malachite. The chemical formula for malachite is **CuCO₃•Cu(OH)₂**.
    (This symbol "•" means that the two compounds are not present in a fixed ratio to each other.)

11. Copper is not corroded by air when it is dry and at normal temperatures. However, in moist air, water and carbon dioxide cause a green coating called verdigris to form on the copper. Verdigris is a form of copper(II) carbonate. The equation is

$$2\ Cu + O_2 + H_2O + CO_2 \longrightarrow CuCO_3 \bullet Cu(OH)_2$$
$$\text{verdigris}$$

Suppose malachite was changed directly to copper in the following reaction:

$$CuCO_3 \bullet Cu(OH)_2 \longrightarrow 2\ Cu + O_2 + H_2O + CO_2$$
$$\text{malachite}$$

What are the differences between the reactions of verdigris and malachite?

10.7 ACTIVITY
EXOTHERMIC AND ENDOTHERMIC PROCESSES

Chemical reactions, physical changes of state, and dissolving processes often involve energy changes. Processes that release energy, such as heat and light, are exothermic processes. Processes that absorb energy are endothermic processes. Temperature changes often occur with these types of processes. Exothermic processes release heat and result in a temperature increase. Endothermic processes absorb heat from the surroundings and result in a temperature decrease of the surroundings.

Even though exothermic chemical reactions produce heat, they sometimes need heat to get them started. For example, in Activity 10.2, magnesium and oxygen produced a significant quantity of heat and light. This reaction is exothermic, but it needed heat to get started.

In an endothermic chemical reaction, energy, such as heat, is needed continuously to keep the reaction going. At other times, there may be a noticeable decrease in temperature as a reaction occurs.

Hot and cold packs are applications of exothermic and endothermic processes. In this activity you will produce a laboratory version of a cold pack and hot pack.

Materials

ammonium nitrate
resealable plastic bags (2)
small plastic bags (2)
calcium chloride

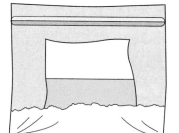

Method

1. Add approximately 200 g of ammonium nitrate to a large resealable plastic bag.
2. Add 100 mL of water to a small plastic bag. Leave the top open.
3. Place the small bag inside the large bag with both openings at the top.
4. Seal the large bag.
5. With the large bag sealed, invert the bags and gently shake. Record your observations.
6. Add approximately 35 g of calcium chloride to the second large bag.
7. Add 100 mL of water to the second small plastic bag. Leave the top open.
8. Place the small bag inside the large bag with both openings at the top.
9. Seal the large bag.
10. With the large bag sealed, invert the bags and gently shake. Record your observations.

Questions and Conclusions

1. What happened to the temperature of the solution as the ammonium nitrate dissolved?
2. What happened to the temperature of the solution as the calcium chloride dissolved?
3. Which process was endothermic and which process was exothermic?
4. What happens to the temperature of the surroundings in an endothermic reaction?

Commercial hot and cold packs

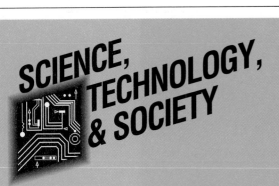

Catalytic Converters

Some chemical reactions occur slowly. Often the rate of a chemical reaction can be increased dramatically by the presence of a catalyst. Catalysts speed up a reaction without being permanently used up in the reaction. They do not appear as a reactant or product.

Since the invention of automobiles, exhaust fumes from cars have contained some unburned hydrocarbons as well as nitrogen monoxide and carbon monoxide. With an efficient and well-tuned engine, fewer of these pollutants are found in the emissions. In order to control pollutants, governments established emission standards for the auto industry.

In 1975 auto manufacturers began installing catalytic converters in cars. These devices convert nitrogen monoxide into nitrogen gas and oxygen gas, and complete the combustion of unburned hydrocarbons and carbon monoxide to carbon dioxide and water vapor.

The catalytic converter looks like a muffler. It contains porous, heat resistant materials coated with a catalyst. The catalyst used in most converters is a mixture of platinum and palladium.

The reactants are absorbed on the surface of the metal, which spreads the reactants over a larger surface area. Catalytic action can be hindered by negative catalysts or inhibitors. These interfere with the surface chemistry of the catalyst. In automobiles, tetraethyl lead in the gasoline has this effect. After 1975 lead-free gasolines were introduced and leaded gasoline was slowly phased out so that today "regular" gasoline means lead-free gasoline.

CAREER OPPORTUNITY

Firefighter

Firefighters are valuable members of every community. They provide an important and sometimes lifesaving service.

They protect property as well as lives and are also well-suited to help in many household emergencies. People in distress from accidents, storms, or floods, as well as fires, often rely on firefighters as their first source of assistance.

Because they have to assist injured people or people who have inhaled smoke, firefighters are trained extensively in first aid and artificial respiration techniques such as cardiopulmonary resuscitation (CPR).

Firefighters have to meet rigorous physical and medical requirements demanding strength and endurance as well as courage and the ability to work as a member of a team. In order to qualify as a firefighter, you will need a high school diploma including courses in mathematics, physics, and chemistry.

Being a firefighter is a rewarding and challenging career and a vital service to the community.

Exercises

Key Words

acid rain
catalyst
combustion
contact process
decomposition
direct combination
endothermic
exothermic
Haber process
spontaneous
    decomposition
synthesis
verdigris

Review

1. What general type of reaction is used in the Haber process?
2. List three uses of
   a. ammonia   b. sulfuric acid
3. Write equations for the three synthesis reactions involved in making sulfuric acid.
4. Calcium carbonate decomposes in the same way as copper(II) carbonate. What is the balanced chemical equation for the decomposition of calcium carbonate?
5. Name the general type of reaction for each of the following:
   a. $Ag(s) + S(g) \longrightarrow AgS(s)$
   b. $2\ H_2O(g) \longrightarrow 2\ H_2(g) + O_2(g)$
   c. $C_3H_8(g) + 5\ O_2(g) \longrightarrow 3\ CO_2(g) + 4\ H_2O(g)$
6. Give two examples each of endothermic and exothermic processes.
7. What happens to the temperature of the surroundings during an endothermic process?

Extension

1. Use the library resource center to find out what property of hydrogen peroxide makes it useful as an antiseptic, bleach, or rocket fuel.
2. Perform an energy analysis in your school or home and determine ten ways to reduce your energy consumption and energy bill.
3. Investigate the efficiency of methods of home heating and report your results to the class.
4. Use the library resource center to research alternative methods of energy production that could reduce energy costs. Present your results to the class.

CHAPTER 11
DISPLACEMENT
REACTIONS

This chapter will introduce you to both single displacement and double displacement reactions. These two types of reactions are two of the four basic patterns that most chemical reactions follow.

In single displacement reactions, reactive metals displace less active metals, or elements, such as hydrogen, from their compounds. Some of these reactions can generate a great deal of energy, producing a molten metal. These reactions are often used in the metallurgy industry in the smelting process.

Reactions that produce precipitates and acid-base neutralization reactions are often double displacement reactions. Many qualitative analysis techniques rely on precipitate-forming double displacement reactions.

A great many applications of single and double displacement reactions are found in the home and industry.

Key Ideas

- Chemical reactions follow one of several general patterns.
- Each type of general reaction pattern has important applications.
- In single displacement reactions the more active metal displaces a less active metal.
- Single and double displacement reactions are often used in metallurgy.
- The metal industry provides many different kinds of jobs.

Molten steel being poured into molds

11.1 ACTIVITY
SINGLE
DISPLACEMENT

Steel being galvanized as it emerges from a bath of molten zinc, which coats the steel in a continuous process

A **single displacement** or substitution reaction fits the general pattern of

$$A + BC \longrightarrow AC + B$$

This type of reaction involves a change in bonding partners. One element displaces or knocks off another element and then substitutes for it.

Zinc is an element that participates in single displacement reactions. It is used to coat objects made of iron to stop them from rusting. This process is called galvanizing

Zinc reacts readily with common acids such as sulfuric acid and hydrochloric acid.

$$Zn + H_2SO_4 \longrightarrow ZnSO_4 + H_2(g)$$

$$Zn + 2\ HCl \longrightarrow ZnCl_2 + H_2(g)$$

Some important zinc compounds are zinc oxide, zinc sulfide, zinc chloride, and zinc sulfate. Zinc oxide is called zinc white and is a white pigment used in artists' oil paints. It is also used in rubber compounding and in sunblock. Zinc sulfide is another white compound used in the production of phosphors used in television screens.

Two zinc compounds that you will make in this activity are zinc chloride and zinc sulfate. Zinc chloride is used in the textile industry and as a wood preservative. Zinc sulfate, known as zinc vitriol, is used to make lithopone paint and is used in dyes in the textile industry.

Materials

burner
test tubes (5)
evaporating dish
copper(II) chloride solution
magnesium ribbon
aluminum turnings
iron turnings (or filings)
mossy zinc
copper(II) sulfate solution

Method

1. Prepare a table to record your observations.
2. Set up four test tubes, each containing 5 mL of copper(II) chloride solution.
3. Add a small piece of magnesium, aluminum, zinc, and iron to each of the four test tubes (only one metal per tube). Record your observations at 5 min intervals.

4. Place a piece of mossy zinc in a test tube and add 5 mL of the copper(II) sulfate solution. Record your observations.

5. Decant the clear liquid from the test tube used in step 4 into an evaporating dish, and evaporate it by heating.

Questions and Conclusions

1. Did all of the test tubes in step 3 show evidence that a reaction was occurring? If not, which test tube(s) did not?

2. Write balanced chemical equations for the reactions in step 3.

3. What is the general pattern for the reactions in step 3?

4. What did you notice about how quickly the reaction took place in each test tube? Rank the metals from most active (reactive) to least active.

5. Which metal was more active, zinc or copper?

6. Fill in the blanks with the word "more" or "less." The ▪▪▪▪ active element displaces the ▪▪▪▪ active element in a single displacement reaction. **Do not write in this textbook.**

11.2
METALS
AND SINGLE
DISPLACEMENT
REACTIONS

Single displacement reactions can be very vigorous and spectacular as the more reactive metal displaces the less active metal.

One example involves the reaction of aluminum powder with iron(III) oxide powder. The more reactive aluminum displaces the iron to form aluminum oxide and produces molten metallic iron. This is called the thermite reaction.

$$2\ Al(s) + Fe_2O_3(s) \longrightarrow Al_2O_3(l) + 2\ Fe(l)$$

The mixture needs to be heated to start the reaction. Once started, large amounts of heat are produced as a byproduct of the reaction. The temperature increase is enough to form molten aluminum oxide and molten iron.

The thermite reaction is used to weld together steel parts, such as railway lines.

Smelting is a method of producing a metal from its ore using a single displacement reaction. The ore, made up of the metallic oxide, is heated with carbon in the form of charcoal or coke. Carbon will displace metals less active than it to form the metal and carbon dioxide.

This method of producing metals was used more than three thousand years ago by the Egyptians in producing copper. Today, smelting is better understood chemically and techniques have improved, but the reaction is still a very useful application of a single displacement reaction.

The thermite reaction produces molten iron.

The production of iron in a blast furnace

Try These

1. In the thermite reaction, which is the more active metal, aluminum or iron? Why?
2. Write the equation for the smelting of copper(II) oxide using charcoal.
3. Aluminum is more active than carbon. Could the smelting process using carbon be used to produce aluminum from aluminum oxide?

11.3 ACTIVITY
METALLIC
CRYSTALS

In Activity 11.1 a metallic element was formed in the single displacement reactions. In this activity the element formed is another metal and it separates from the reaction in an interesting way.

Materials

100 mL beaker

250 mL beakers (2)

watch glasses (2)

pen

15 cm piece of copper wire

silver nitrate solution

zinc(II) sulfate solution

copper(II) sulfate solution

zinc strip

sandpaper

copper strip

Caution: Silver nitrate is corrosive and causes burns. Handle it with care. Wear goggles and an apron. If you spill any on yourself, rinse it off immediately with large amounts of cold water and have your partner tell your teacher.

Method

Part A

1. Use the sandpaper to clean the copper wire. Wind the copper wire into a coil around a pen.
2. Pour 50 mL of the silver nitrate solution into a 100 mL beaker.
3. Place the wire into the solution so that the coil is immersed and the end of the wire is bent over the lip of the beaker.
4. Cover the beaker with a watch glass and set it where it can stay undisturbed until the next class. At that time, examine the results and record your observations.

Part B

1. Use the sandpaper to clean the zinc strip and copper strip.
2. Pour 150 mL of copper(II) sulfate solution into a 250 mL beaker.
3. Place a clean strip of zinc metal into the solution.
4. Repeat steps 2 and 3 with zinc sulfate solution and a copper strip.
5. Observe any changes that occur.
6. Label your beakers and let the reactions continue overnight.
7. Record your observations.

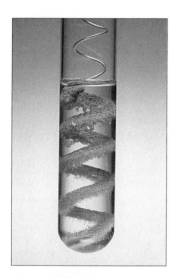

Silver crystals being formed on a copper helix. The blue color of the solution shows that copper ions are present.

Questions and Conclusions

1. Write the word equation for each single displacement reaction that occurred.
2. What is the general pattern of the reactions?
3. Which element in each of the following pairs displaced the other from the compound?
   a. copper or silver   b. zinc or copper
4. In Part B, which metal appears to be most active, zinc or copper? Explain why.

11.4 TREATMENT OF METALS

Galvanizing an iron object is one way to protect it from rusting. Another way is to coat it with paint.

Aside from protecting metal from corrosion, there are other ways to change metals to make them better suited for specific purposes. Steel is an alloy produced primarily from iron and carbon. An alloy is a substance that has metallic properties and contains more than one element.

Pure iron is malleable, soft, and ductile. Wrought iron, which is a very pure form of iron, is used in ornamental ironwork because it can be bent easily into intricate shapes. Steel is used for such things as pipes, wire products, and structural steel in buildings, bridges, and ocean liners.

The Rainbow Bridge, spanning the Niagara River between Canada and the United States, is reinforced with steel.

Large ships are made from the alloy steel.

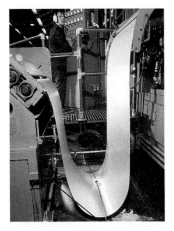

The malleability of metals allows them to be rolled into thin sheets. This property is used in industry to manufacture sheet metal for a wide variety of applications.

Steel is made from pig iron. This is an impure form of iron produced in a blast furnace. Pig iron contains up to 5% carbon and smaller amounts of impurities, such as phosphorus and sulfur.

Steel is made by removing most of the impurities in pig iron. The amount of carbon in the steel gives it certain properties. Steel generally contains less than 2% carbon. For malleable steel used in auto body parts, the carbon level is about 0.2%. Tougher steel has a higher carbon level.

Try These

1. a. What is an alloy?
   b. Why are alloys used?
2. What are the advantages of steel over pure iron?
3. How does the percentage of carbon present in steel affect its malleability?

SCIENCE & TECHNOLOGY

Would You Like That in Metal or Plastic?

Research in organic chemistry has led to the widespread use of plastics as substitutes for metals. Some common examples are plumbing pipes, auto body panels, and snow shovels.

Over the last fifty years, production of plastics has increased dramatically: over sixty-five million tonnes are produced worldwide every year. One great advantage of plastics is that they can be made to have a wide variety of properties, including strength, flexibility, rigidity, resistance to chemicals, and so on, at low cost.

Plastics are made by polymerization. This can be done either by adding units together when double bonds are broken or by reacting monomers together.

As with other substances, the properties of polymers are related to their structure.

Thermoplastics consist of long tangled chains held together by weak forces between the chains.

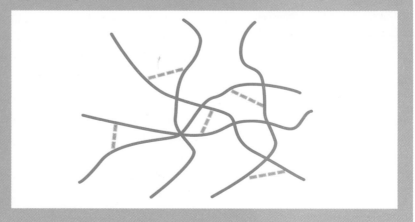

These plastics stretch easily, soften when heated, and have low melting points. Examples of these are nylon, polypropylene, and polystyrene. What are some uses of each of these?

Thermoset plastics consist of large cross-linked chains of molecules held together by strong forces between the chains which prevent the chains from moving relative to each other. Examples of these plastics are phenolic and melamine resins and epoxies.

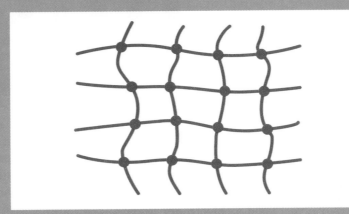

What type of plastic, thermoset or thermoplastic, would be most suitable for auto body parts? Why?

11.5 MORE COMMERCIAL APPLICATIONS OF METALLURGY

Each of the types of reactions studied in the previous activities involved a metal or metallic compound. Two of these reactions are very important to the metal industry.

Metallurgy is the science of working with metals. The United States and Canada are both world leaders in mineral production and exports. A large segment of the economy depends on mining iron and making steel and steel products.

Metals are extracted from their ores by certain chemical reactions. Aluminum is the most abundant metal found in the earth's crust, but it is always found bonded to other elements rather than as the pure metal. It makes up 8% of the earth's crust and is found in minerals such as bauxite, mica, cryolite, and some clays. The main commercial ore is bauxite $(Al_2O_3 \cdot 2 \, H_2O)$. (The "•" indicates that Al_2O_3 is loosely associated water in a 1 : 2 ratio.)

Because aluminum is more reactive than carbon, it cannot be smelted from its ore in the same way as iron or copper. Single displacement reactions that could produce aluminum were tried, but the cost was too high to make them commercially feasible.

Metallurgy Technician

Metallurgy is the science of working with metals. A metallurgy technician's responsibility is to ensure that metal products meet specific standards.

One of the duties of a metallurgy technician may be to see how well metals stand up to stress. This metal stress can occur in a jet turbine blade, an alloy used in solder, or steel used in kitchen knives.

The metallurgy technician's work often involves looking for defects in metals or processing materials. In most cases, a technician applies his or her knowledge to a particular process or problem. As with other technician careers, interested students should enjoy science and technical work. They should also be comfortable working with numbers and equipment. Since writing reports is often part of the job, metallurgy technicians must be able to express themselves in writing.

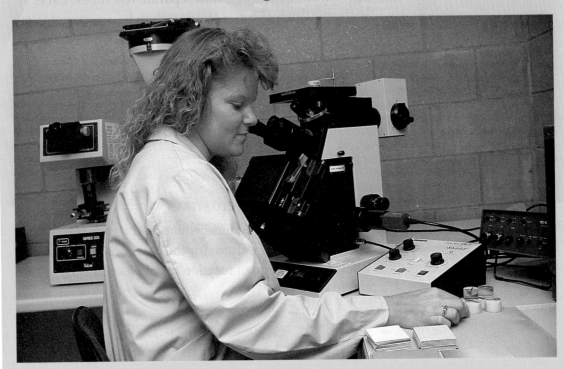

Exercises

Key Words

alloy

bauxite

double displacement

electrolysis

galvanizing

Hall-Héroult process

metallurgy

roasting

single displacement

smelting

thermoplastics

thermoset plastics

Review

1. Name three important uses of the element zinc.
2. a. What compound is called zinc white?
   b. What are two uses of zinc white?
3. a. Write the equation for the thermite reaction.
   b. What type of reaction is the thermite reaction?
   c. Give one application of the thermite reaction.
4. Define "smelting."
5. a. List three metals that can be produced by smelting.
   b. How does the reactivity of metals produced by smelting compare to the reactivity of carbon?
6. What general type of reaction occurs when a metal is produced by electrolysis?
7. What pollutants are produced by the roasting of metal sulfide ores?
8. Name the general type of reaction for each of the following:
   a. $Zn(s) + CuSO_4(aq) \longrightarrow ZnSO_4(aq) + Cu(s)$
   b. $NaOH(aq) + HCl(aq) \longrightarrow NaCl(aq) + H_2O(l)$

Extension

1. Write a report on the alloys of steel and their uses in society. Present your report to the class.
2. Prepare a poster for class display illustrating the composition, manufacture, and uses of one alloy.
3. a. What role does sulfur dioxide play in the formation of acid rain?
   b. Use the library resource center to find out the effects of acid rain on the following:
      i. lakes   ii. forests   iii. human beings
   c. Outline the types of legislation and agreements on acid rain that are in effect between Canada and the United States. Why might it be difficult to get agreement between the two countries?
4. Use the library resource center to research the use of chemical reactions to recover expensive metals from wastes or scrap.

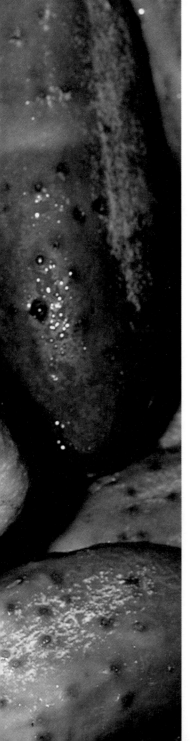

CHAPTER 12
PROPERTIES OF ACIDS AND BASES

In previous chapters you learned how to classify matter. First you classified matter into elements, mixtures, and compounds. Then you classified elements into metals and nonmetals. In this chapter you will study one of the oldest classification systems of matter, acids, and bases.

Key Ideas

- Compounds classified as acids have characteristic properties.
- Compounds classified as bases have characteristic properties, some of which are quite different from those of acids.
- Acids and bases are corrosive and must be handled with extreme caution.
- Acids and bases can be recognized by their names.

12.1
INTRODUCTION TO ACIDS AND BASES

Almost everyone has experienced biting into a sour apple or sour grapes. The word "sour" is used to describe a particular taste. This is also where the expression "acid tongue" comes from when used to describe a person who makes sharp remarks. The first class of substances was those that had a sour taste. They were called acids. The word **acid** comes from the Latin word *acidus*, which means sour.

Bases are the second class of compounds and perhaps are not as familiar to you. **Bases** react with acids to form new products. You probably are more familiar with the term **"antacid"** than you are with the term "base," which is the same thing. A base has a bitter taste. Have you ever accidentally tasted soap or shampoo? If you did, you have experienced a bitter taste. What else could you classify as being bitter?

12.2   ACTIVITY
SOUR VS. BITTER

In this activity you will classify different foods according to taste. If you have food allergies, substitutions can be made.

Apples can be sour.

Materials

apple	milk
lemon	tonic water
Brussels sprouts	vinegar
dill pickles	

Method

1. Obtain a sample of each of the foods listed.
2. Prepare a chart with the headings sour and bitter.
3. Taste each sample and classify it as sour or bitter.
4. Record your observations.

Questions and Conclusions

1. What class of compounds is sour?
2. What class of compounds is bitter?
3. Is taste a good method for identifying acids and bases? Explain.
4. Use the library resource center to find out the names of the acids and bases present in each of the samples.

12.3 ACTIVITY PROPERTIES OF ACIDS

Many acids taste sour. Taste-testing, however, should be done very cautiously and only with the permission of your teacher.

In this activity you will be using a much safer method of identifying acids and bases than taste-testing. You will use litmus paper. Litmus is a dye that is extracted from a type of plant called lichen. Litmus is blue in its natural state and turns red if an acid is added to it.

Materials

test tubes (2)
glass slides (2)
glass stirring rods (2)
medicine dropper
magnesium ribbon
red and blue litmus paper
dilute acetic acid
dilute hydrochloric acid
phenolphthalein indicator

Caution: Acids are corrosive and cause burns. Handle them with care. Wear goggles and an apron. If you spill any on yourself, rinse it off immediately with large amounts of cold water and have your partner tell your teacher.

Be careful when handling acids and bases.

Method

1. Prepare a data chart like the one shown below.

	Hydrochloric Acid	Acetic Acid
red litmus paper		
blue litmus paper		
phenolphthalein		
magnesium		
touch test	DO NOT TOUCH.	

2. Label one test tube 1 and the other 2.
3. Put 4 mL of water into each of the two test tubes.
4. Add 15 drops of hydrochloric acid to the first test tube, using the medicine dropper.
5. Add 15 drops of acetic acid to the second test tube.
6. Mix the contents of each test tube using different stirring rods.
7. Use the rod to add a drop of the hydrochloric acid solution to
    a. a strip of red litmus paper
    b. a strip of blue litmus paper
    c. a drop of phenolphthalein indicator on a glass slide
8. Record your results in your chart.
9. Rinse the glass rod and use it to add a drop of the acetic acid solution to
    a. a strip of red litmus paper
    b. a strip of blue litmus paper
    c. a drop of phenolphthalein indicator on a glass slide
10. Record your results.
11. Add a piece of magnesium ribbon to each of the test tubes of hydrochloric and acetic acid. Record your observations.
12. What gas do you think was produced in step 11? Carry out a test to see if your prediction was correct.
13. Place two drops of dilute acetic acid on your little finger with a stirring rod. Describe how it feels.

Questions and Conclusions

1. List three general properties of an acid.
2. What effect does an acid have on a metal?
3. In what household products would you find the following?
    a. hydrochloric acid
    b. acetic acid

12.4 ACTIVITY
PROPERTIES
OF BASES

In this activity you will investigate the properties of two bases: sodium hydroxide and ammonium hydroxide.

Materials

test tubes (2)
glass stirring rods (2)
glass slides (2)
medicine dropper
red and blue litmus paper
magnesium ribbon
dilute sodium hydroxide solution
ammonium hydroxide solution
phenolphthalein indicator

Caution: Bases are corrosive and cause burns. Handle them with care. Wear goggles and an apron. If you spill any on yourself, rinse it off immediately with large amounts of cold water and have your partner tell your teacher.

Method

1. Prepare a chart for your data similar to the one in Activity 12.3. Use it to record your observations.
2. Label one test tube 1 and the other test tube 2. Put 4 mL of water into each of the test tubes.
3. Add 15 drops of sodium hydroxide to the first test tube using the medicine dropper.
4. Add 15 drops of ammonium hydroxide to the second test tube.
5. Stir each of the solutions.
6. Use a glass rod to add a drop of sodium hydroxide to
   a. red litmus paper
   b. blue litmus paper
   c. a drop of phenolphthalein indicator on a slide
7. Use a glass rod to add a drop of ammonium hydroxide to
   a. red litmus paper
   b. blue litmus paper
   c. a drop of phenolphthalein indicator on a slide

8. Add a piece of magnesium ribbon to each of the sodium hydroxide and ammonium hydroxide solutions.

9. Use a medicine dropper to put a drop of dilute sodium hydroxide solution between two of your fingers. Describe how it feels.

Questions and Conclusions

1. List three general properties of a base.
2. Why do all bases possess common properties? (Hint: What do the names of all bases have in common?)
3. In which common household products would you find
   a. sodium hydroxide?
   b. ammonium hydroxide?
4. Are acids and bases equally dangerous? Explain.

12.5
PROPERTIES
OF ACIDS
AND BASES

Early scientists classified substances in groups if they had similar properties. The group of substances that are called acids produce solutions that

1. conduct electricity
2. react with active metals to produce hydrogen gas
3. change the color of litmus and other dyes
4. react with bases to form new substances
5. have a sour taste

Figure 12.1   Many common products contain acids.

The group of substances that are called bases produce solutions that

1. conduct electricity
2. change the color of litmus and other dyes to different colors from those of acid solutions
3. react with acids to produce new substances
4. are slippery to the touch
5. have a bitter taste

Figure 12.2   Many common products contain bases.

12.6 ACTIVITY
ACID MEETS BASE—
A CASE FOR NEUTRALIZATION

In this activity you will observe a neutralization reaction. The quantitative method you will use is called a titration.

Materials

retort stand	glass slide
inoculation loop	graduated cylinder
wire gauze	sodium hydroxide solution
ring clamp	hydrochloric acid solution
burner	silver nitrate solution
evaporating dish	phenolphthalein indicator
medicine dropper	disposable rubber gloves
glass stirring rod	

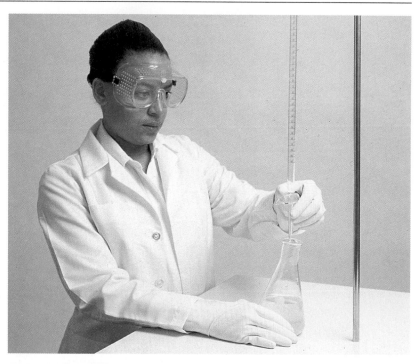

Wear goggles, aprons, and gloves when performing titration experiments.

Caution: Acids, bases, and silver nitrate are corrosive and cause burns. Handle them with care. Wear goggles and an apron. If you spill any on yourself, rinse it off immediately with large amounts of cold water and have your partner tell your teacher.

Method

1. To find the volume that your medicine dropper will hold, fill the dropper with water and empty it drop by drop into a graduated cylinder. Continue filling and emptying the dropper until you have placed 100 drops into the graduated cylinder.
2. Read the volume on the cylinder.
3. Calculate the average volume of one drop. An example is as follows:

   Number of drops   = 100

   If the volume of water in
   the graduated cylinder   = 10 mL

   Then 1 drop of water   = 10 mL ÷ 100 drops

                            = 0.1 mL/drop

   The volume of one drop of water would be 0.1 mL.
4. Empty the graduated cylinder.

5. Measure 5 mL of the sodium hydroxide solution and pour it into an evaporating dish.
6. Add two drops of phenolphthalein indicator.
7. Record your observations.
8. Rinse your dropper with the hydrochloric acid solution and then fill it with the acid.
9. Add acid one drop at a time to the solution in the evaporating dish, stirring after each addition.
10. Continue to add acid, keeping track of the number of drops added.
11. Count the total number of drops of acid needed to make the solution colorless.
12. Calculate the volume of acid added. Use the volume per drop that you calculated in step 3.
13. Add two drops of the solution to one drop of silver nitrate solution on a glass slide. Observe.
14. Place the evaporating dish on a wire gauze and ring clamp attached to a retort stand.
15. Dip an inoculation loop into the solution in the evaporating dish.
16. Hold the loop in a burner flame. Observe the color of the flame.
17. Heat the solution in the evaporating dish.
18. Remove the evaporating dish from the heat and allow it to cool.
19. Describe the residue.

Questions and Conclusions

1. State the reason for the color change noted in step 7.
2. How did the volume of acid added compare with the volume of base taken?
3. Why was the phenolphthalein indicator added?
4. What ion does silver nitrate test for in step 13?
5. What color flame did you observe in step 16?
6. What metal ion was present in the solution?
7. Describe the residue. What is the residue?
8. Write the word equation for the reaction.
9. Name the type of reaction.
10. What do you predict the pH to be? Explain why.
11. How would your results change if a more dilute solution of base than acid was used?

12.7
NEUTRALIZATION

When acids and bases are combined, they neutralize each other. This means that two new products are formed that are neutral compounds. These products are a salt and water.

Millions of dollars are spent each year by the public on bases that relieve excess stomach acid. The reaction that occurs is between the hydrogen ions in the stomach's acid and the hydroxide ions in the antacid.

When the sodium hydroxide breaks down in solution, it forms sodium ions and hydroxide ions.

$$NaOH(aq) \longrightarrow Na^+ + OH^-(aq)$$

When hydrochloric acid is dissolved in water, it forms hydrogen ions and chloride ions.

$$HCl(aq) \longrightarrow H^+(aq) + Cl^-(aq)$$

When an acid solution is added to a basic solution, the hydrogen ions combine with the hydroxide ions to form water and the metal ion combines with the nonmetal ion to form a salt, as shown in Figure 12.3.

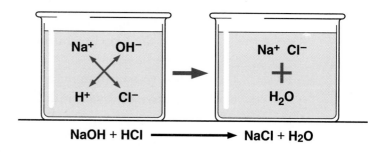

Figure 12.3   The formation of salt and water in a neutralization reaction

Try These

1. What two products are formed when an acid and a base react?
2. What ions are actually responsible for the reaction?
3. Which of the four types of reactions would a neutralization reaction come under?
4. Write a word equation and a balanced chemical equation for the reaction of calcium hydroxide with hydrochloric acid. Name the products.

Upset Stomach, Heartburn, and Antacids

Our stomachs digest foods using gastric juices, including hydrochloric acid. The stomach itself is protected from the hydrochloric acid by a mucus coating.

The esophagus connects our mouth to our stomach. Because digestion does not occur in the esophagus, there is no protective coating. Occasionally stomach acids are splashed up into the esophagus. The irritation caused is called "heartburn," although it is actually in the esophagus.

To relieve this sensation, people take antacids or bases to neutralize the acid. Many antacids contain a hydroxide base such as magnesium hydroxide, $Mg(OH)_2$, or aluminum hydroxide, $Al(OH)_3$. The neutralization reactions are

$$Mg(OH)_2 + 2\ HCl \longrightarrow MgCl_2 + 2\ H_2O$$

or

$$Al(OH)_3 + 3\ HCl \longrightarrow AlCl_3 + 3\ H_2O$$

Magnesium hydroxide is a laxative, but aluminum hydroxide alleviates diarrhea. Some antacids contain a mixture of both in an attempt to balance the side effects.

Some antacids use bicarbonate and carbonates to neutralize stomach acid. Examples of these reactions are

$$NaHCO_3 + HCl \longrightarrow NaCl + H_2O + CO_2$$

or

$$MgCO_3 + 2\ HCl \longrightarrow MgCl_2 + H_2O + CO_2$$

These antacids produce carbon dioxide as they react with stomach acid.

12.8 ACTIVITY
MAKING YOUR OWN ACIDS AND BASES

In this activity you will be producing oxides and then making acids and bases from them. Oxides are compounds that are formed when an element combines with oxygen. The oxides you will be producing are oxides of magnesium, calcium, sulfur, and carbon. Oxides of metals produce basic solutions in water. Oxides of nonmetals produce acidic solutions in water.

Materials

burner	litmus paper
metal tongs	calcium
deflagrating spoon	carbon
gas bottles (3)	sulfur
watch glass	magnesium ribbon
test tube	oxygen gas
glass stirring rod	cobalt blue glass shield

Caution: Goggles should be worn at all times during this activity. Do not look directly at the light produced by burning magnesium. It can cause eye damage. A cobalt blue glass shield should be used to view the reaction. The sulfur reaction (step 8) should be carried out in a fume hood.

Method

1. Prepare a chart as shown below and record all of your observations in it.

Element	Description of Element	Description of Product	Description of Litmus in Solution	Formula of Oxide
calcium				
carbon				
magnesium				
sulfur				

2. Put your sample of calcium in a deflagrating spoon. Record a description of calcium.
3. Heat the calcium briefly in a burner flame. Once it begins to burn, remove it from the flame.
4. Lower the spoon into a gas bottle of oxygen and cover the opening with a watch glass.
5. Once the calcium has finished burning, describe the product.

6. Pour 5 mL of water into the gas bottle and stir.
7. Test your solution with litmus.
8. Repeat steps 2–7 using carbon and sulfur.
9. Using metal tongs, hold a 2-cm strip of magnesium in a burner flame.
10. Add the product to 4 mL of water in the test tube and stir. Test the solution with litmus paper.

Questions and Conclusions

1. Which of the dissolved oxides changed the color of the red litmus paper?
2. Which of the dissolved oxides did not change the color of the litmus?
3. Write balanced chemical equations for the combustion of calcium, carbon, sulfur, and magnesium.
4. Name the four oxides formed from the elements named in question 3.
5. What generalization can you make regarding the solutions formed from metallic and nonmetallic oxides?

12.9
WHAT'S
IN A NAME?

There are two large groups of acids. Acids that contain two elements are called binary acids. Acids that contain more than two elements are called oxyacids.

The binary compound that results when hydrogen and chlorine are combined is called hydrogen chloride unless it is dissolved in water. When it is dissolved in water, it displays acidic properties and turns blue litmus red. These properties are quite different from those of the original compound. Such compounds dissolved in water are called acids. The rules for naming these binary acids are as follows:

1. Use **hydro** as the prefix.
2. Then add the stem name of the second element and add an **ic** ending.

For example, hydrogen chloride dissolved in water forms the binary acid **hydro**chloric acid.

Table 12.1   Names of common binary acids

Formula	Name of Pure Compound	Name of Acid
HCl	hydrogen chloride	**hydro**chloric acid
HF	hydrogen fluoride	**hydro**fluoric acid
HBr	hydrogen bromide	**hydro**bromic acid

When a binary compound composed of oxygen and another nonmetal reacts with water, a second type of acid is formed. This type of acid is called an oxyacid.

Different oxyacids often contain the same elements but have different numbers of oxygen atoms present. An example of this is H_2SO_3 and H_2SO_4.

Table 12.2   Five common oxyacids

Formula	Name of Pure Compound	Name of Acid
H_3PO_4	trihydrogen phosphate	phosphoric acid
H_2SO_4	dihydrogen sulfate	sulfuric acid
H_2CO_3	dihydrogen carbonate	carbonic acid
$HClO_3$	hydrogen chlorate	chloric acid
HNO_3	hydrogen nitrate	nitric acid

Different oxyacids can be formed by changing the number of oxygen atoms present. This is where the following naming system comes in handy.

1. If the acid contains one more oxygen atom than the common acid, the prefix **per** is added to the name of the oxyacid.
2. If the acid contains one less oxygen atom than the common acid, the ending is changed from **ic** to **ous**.
3. If the acid contains two less oxygen atoms than the common acid, the prefix **hypo** is added and the ending **ous** is used.

The following are all the possible oxyacids for $HClO_3$:

$HClO_4$		**per**chloric acid
$HClO_3$	common acid	**chloric acid**
$HClO_2$		chlor**ous** acid
$HClO$		**hypo**chlor**ous** acid

You can tell a base by its name. It usually starts with the name of a metal and ends with hydroxide. For example,

(metal) (hydroxide)

NaOH

sodium hydroxide

Try These

1. Name the following oxyacids:
   a. H_3PO_4 c. HNO_2 e. H_2CO_3
   b. $HClO_2$ d. H_3PO_3 f. HNO_3
2. Name the following compounds:
   a. $Ca(OH)_2$ c. $Al(OH)_3$ e. $LiOH$
   b. $Ba(OH)_2$ d. $RbOH$ f. $Mg(OH)_2$

Exercises

Key Words

acid

antacid

base

binary acid

hydroxide

litmus

neutralize

oxide

oxyacid

Review

1. Define the terms "acid" and "base."
2. List five characteristics of an acid. List five characteristics of a base.
3. List two safety precautions taken when handling acids and bases.
4. What plant is litmus derived from?
5. You are given an unknown solution and asked to classify it as basic or acidic. What two tests would you perform?
6. Write a word equation for zinc reacting with hydrochloric acid.
7. What type of oxides are base producers? acid producers?
8. Give the names and formulas of the five common oxyacids.
9. What is the difference between an oxyacid and a binary acid?
10. Give two examples of a binary acid.
11. What ending do all bases have in common?
12. What do the prefixes "per" and "hypo" mean?
13. An acid has two less oxygen atoms than its common oxyacid. What changes would you make to the oxyacids name?
14. How would you make a base in a lab?
15. Give the names and formulas of five bases.

Extension

1. Acid precipitation is the product of non-metal oxides reacting with moisture in the air. Research the sources of these non-metal oxides and explain what is being done to eliminate the problem.
2. Go to a pharmacy and list the price, number of tablets, and active ingredients of at least three antacids. Rank them in order of decreasing value.

CHAPTER 13
INDICATORS AND pH

Acids and bases are used extensively in our lives. Acids such as vinegar, and bases such as lime or baking soda are commonly encountered each day. Chemists use different methods to designate a substance acidic or basic.

The first method is based on how the solution reacts and is qualitative. The second method is based on a number and is done quantitatively. In this chapter you will be examining both qualitative methods and quantitative methods used in working with acids and bases.

Key Ideas

- Indicators show the presence of an acid or base by changing colors.
- Acids and bases are found widely in both nature and household products.
- The pH scale is used to measure acidity and basicity.

Red cabbage can be used as an indicator.

13.1
INDICATORS

An indicator is used to tell you about a certain condition. For example, dark heavy clouds may *indicate* that a storm is coming. The fuel gauge in a car *indicates* how much fuel is left in the tank. The clouds and the gas gauge are both indicators.

Dark clouds may *indicate* that a storm is on the way.

In a laboratory chemical indicators are used. These indicators show the acidic or basic condition of a solution by a change in color. An indicator is a substance that reacts with an acid or base to show a definite color change.

Try These

1. What is an indicator?
2. Give an example of a chemical indicator.
3. What does a change in the color of an indicator tell you about a reaction?

13.2 ACTIVITY
ACIDS AND BASES
AT HOME

Acids or bases are found in many places in the home. They may be in the medicine chest, laundry room, kitchen cupboard, or bathroom cupboard. In this activity you will prepare some standard acid and base solutions. You will then use cabbage juice as a colored indicator of acid or base to create standards of acidity and basicity. The pH of common substances can be determined by comparing them with these standards. A substance with a pH of 7 is neutral. Less than pH 7 is acidic and greater than pH 7 is basic.

Materials

red cabbage
lemon
white vinegar
boric acid solution
sodium bicarbonate
    (sodium hydrogen carbonate) solution
borax solution
washing soda solution
drain cleaner
100 mL beaker
250 mL beaker
test tubes (8)
burner or hot plate
unknown solution
glass stirring rod

Caution: Drain cleaner is corrosive and causes burns.
Handle it with care. Wear goggles and an apron.
If you spill any on yourself, rinse it off immediately with
large amounts of cold water and have your partner
tell your teacher.

Method

1. Label the test tubes as follows: pH 2, pH 3, pH 5, pH 7, pH 8, pH 9, pH 12, and pH 14.
2. Place 75 mL of shredded red cabbage in a 250 mL beaker and add 100 mL of water.
3. Gently heat the solution until it boils. Continue to heat until the solution becomes a deep purple. Allow the solution to cool.
4. Pour 2 mL of the cabbage indicator into each of the eight test tubes.
5. Add 10 mL of each of the following solutions to each test tube as listed below.
    Test tube 1—freshly squeezed lemon juice
    Test tube 2—white vinegar
    Test tube 3—boric acid solution
    Test tube 4—water

Test tube 5—sodium bicarbonate solution

Test tube 6—borax solution

Test tube 7—washing soda

Test tube 8—drain cleaner

6. Stir each of the eight solutions with a stirring rod. Rinse the rod between test tubes.

7. Record the color change observed in test tubes 1–8.

8. Place 10 mL of your unknown solution into a test tube. Add 2 mL of the prepared indicator solution to the test tube and stir with a stirring rod.

9. Determine the approximate pH by comparing the color of the solution to that of your standards.

10. Select one of the following products and determine its pH:

a. shampoo   d. baking powder

b. ammonia   e. fertilizer

c. juice

Questions and Conclusions

1. Describe how the color of an indicator changes when an acidic solution is added and when a basic solution is added.

2. What color did the cabbage juice become in the following pH ranges?

a. pH 1–3   c. pH 8–10

b. pH 4–6   d. pH >12

3. What foods other than red cabbage might serve as acid-base indicators? If possible, try them with the materials in this activity.

13.3 ACTIVITY

TESTING FOR ACIDS AND BASES IN TEN CONSUMER PRODUCTS

Many common household products contain either acids or bases. Each product has its own chemical properties that make it useful for different chores. When you have a tough cut of beef, you can make it tender by marinating it. The marinade, whether it is lemon juice, wine, or vinegar, reacts with the tough connective tissue to soften it. When you clean your roasting pan after cooking the beef, you will most likely use soap. The soap contains a base that reacts with the oils in the pan to dissolve them.

In this activity you will be testing for the presence of acids and bases in ten different products.

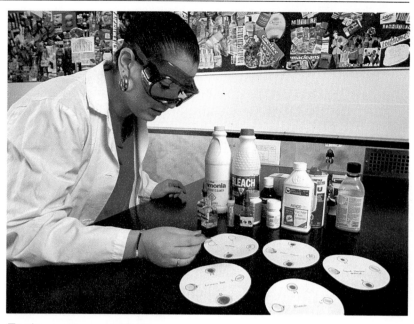

Testing consumer products

Materials

medicine dropper

filter papers (10)

antacid tablet

toilet bowl cleaner

lemon juice

coffee

vinegar

liquid stomach antacid

drain cleaner

window cleaner

bleach

tea

indicators:

    methyl orange

    phenolphthalein

    litmus

    bromthymol blue

Caution: Drain cleaner, toilet bowl cleaner, and bleach are corrosive and cause burns. Handle them with care. Wear goggles and an apron. If you spill any on yourself, rinse it off immediately with large amounts of cold water and have your partner tell your teacher.

Do not mix any of the consumer products with each other. A toxic chemical could be produced.

Method

1. Prepare a chart to summarize your results.

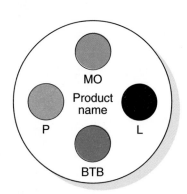

Key:
MO = Methyl orange
P   = Phenolphthalein
L   = Litmus
BTB = Bromthymol blue

2. Prepare ten pieces of filter paper as shown above.
3. Place a drop of each indicator to be used at the edge of the paper.
4. Write the name of the indicator below each drop.
5. Write the name of each product being tested in the center of the piece of filter paper.
6. Allow the filter paper to dry.
7. Use a medicine dropper to add a sample of the product being tested to each indicator. Rinse the dropper between products.
8. Record any color changes.

Questions and Conclusions

1. Use the chart below to determine whether the product contained an acid or a base.

Indicator	Color with Acid	Color with Base
methyl orange	red	yellow
litmus	red	blue
bromthymol blue	yellow	blue
phenolphthalein	colorless	pink

2. Find out the name of the acid present in each of the products containing acid.
3. Find out the name of the base present in each of the products containing base.

13.4
THE RELATIVE STRENGTH OF ACIDS AND BASES

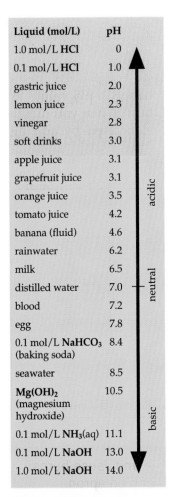

Liquid (mol/L)	pH
1.0 mol/L **HCl**	0
0.1 mol/L **HCl**	1.0
gastric juice	2.0
lemon juice	2.3
vinegar	2.8
soft drinks	3.0
apple juice	3.1
grapefruit juice	3.1
orange juice	3.5
tomato juice	4.2
banana (fluid)	4.6
rainwater	6.2
milk	6.5
distilled water	7.0
blood	7.2
egg	7.8
0.1 mol/L **NaHCO₃** (baking soda)	8.4
seawater	8.5
Mg(OH)₂ (magnesium hydroxide)	10.5
0.1 mol/L **NH₃**(aq)	11.1
0.1 mol/L **NaOH**	13.0
1.0 mol/L **NaOH**	14.0

acidic

neutral

basic

Figure 13.1
The pH of some common substances

In Activity 13.3 you learned that an indicator can tell you whether a substance is an acid or a base. Indicators can also be used to tell us how strong an acid or base is. The strength of the acid or base determines the color to which the indicator will change.

Some acids are very corrosive. For example, battery acid can burn a hole through your clothing in seconds. However, a doctor may use a very weak acid to bathe a patient's infected eye.

Some strong bases are very corrosive. For example, toilet bowl cleaner is very dangerous if spilled on your skin. At the same time, we use a weak base to relieve an acidic stomach.

The strength of an acid or a base can be measured using the pH scale. The pH scale indicates how much acidity is present. Acidity is caused by the presence of free hydrogen ions. The stinging you feel if lemon juice is placed on a cut is due to the H^+ ions present.

The pH scale has a range from 0–14. The closer the pH of a substance is to zero, the more acidic it is. The closer the pH of a substance is to 14, the more basic it is. Pure water has a pH of 7 and is considered to be neutral. We will take a closer look at how this pH is determined.

When water is in the liquid state it exists as a combination of hydrogen and hydroxide ions:

$$H_2O \longrightarrow H^+ + OH^-$$

Water is considered neutral because it contains equal numbers of hydrogen ions and hydroxide ions. The hydrogen ion concentration, $[H^+]$, equals the hydroxide ion concentration $[OH^-]$. (Square brackets around a chemical symbol stand for the words "concentration of.")

As the pH decreases from seven to zero, it indicates that the substance is increasingly more acidic. The hydrogen and hydroxide ion concentrations are no longer equal. The hydrogen ion concentration in an acidic substance is always greater than the hydroxide ion concentration.

As the pH increases from 7 to 14, the substance becomes increasingly more basic. The hydroxide ion concentration in a basic substance is always greater than the hydrogen ion concentration.

Table 13.1   The relative concentration of hydrogen and hydroxide ions

Type of Substance	Concentration [ ]	pH
acidic	$[H^+] > [OH^-]$	less than 7
neutral	$[H^+] = [OH^-]$	equal to 7
basic	$[H^+] < [OH^-]$	more than 7

SCIENCE & TECHNOLOGY

Blood cells

Maintaining Blood pH

Oxygen, which is vital to life, is carried through our blood. The chemical responsible for transporting oxygen is the hemoglobin molecule. The amount of oxygen that can be carried by hemoglobin varies with pH level. The blood pH has to be maintained within a very narrow range.

Normal blood pH at room temperature is 7.35. If the blood pH falls below 7.2, a condition called acidosis sets in. This can be caused by starvation, emphysema, diabetes, or an overdose of aspirin or narcotics. Acidosis is a serious condition.

Similarly, if the blood pH goes above 7.5, alkalosis occurs. This condition can be brought on by continual vomiting and hysterics, kidney disease, or bicarbonate overdose.

Fortunately, our blood has a mechanism to counteract changes in pH. It is a carbonate buffer system. In this system, acidosis is prevented by a bicarbonate ion reacting with $H^+(aq)$:

$$HCO_3^-(aq) + H^+(aq) \longrightarrow H_2CO_3(aq)$$

Alkalosis is prevented by either CO_2 or H_2CO_3 reacting with any hydroxide ions:

$$H_2CO_3(aq) + OH^-(aq) \longrightarrow HCO_3^-(aq) + H_2O$$

The carbonate buffer system, which includes the bicarbonate ion, dissolved carbon dioxide, and carbonic acid, is vital to proper blood chemistry.

13.6 ACTIVITY

QUANTITATIVE TESTS OF THE STRENGTH OF SOME COMMON ACIDS AND BASES

In this activity you will investigate the relative acidic or basic strength of some household products using a quantitative technique. You also will be assessing the warning labels on each product.

Table 13.2   Relative strengths of some acids and bases

Chemical Name	Formula	Common Name	Strength
sulfuric acid	H_2SO_4	battery acid	strong
hydrofluoric acid	HF	etching acid	strong
citric acid	$H_3C_6H_5O_7$	lemon juice	weak
sodium hydroxide	NaOH	caustic soda	strong
calcium hydroxide	$Ca(OH)_2$	lime	weak

Materials

universal indicator
  or pH meter
soap
drain cleaner
vinegar

distilled water
baking soda
cola
lemon juice

Caution: Acids and bases are corrosive and cause burns. Handle them with care. Wear goggles and an apron.
If you spill any on yourself, rinse it off immediately with large amounts of cold water and have your partner tell your teacher.

Method

1. Prepare a data chart listing all the materials you will be testing.
2. Use a pH meter or universal indicator to find the pH of each product.
3. Record each pH in your chart.
4. Arrange the products in increasing order of acidity.

Questions and Conclusions

1. Which products tested were strongly acidic, strongly basic, weakly acidic, weakly basic?
2. Which of the materials tested should carry a warning label? What should the label read?

CAREER OPPORTUNITY

Water-test Laboratory Technician

Water-test laboratory technicians may work for a branch of the federal government or local government, or for private environmental testing companies. They perform a wide variety of routine and technical chemical tests. The samples they test may be brought to them by field technicians or by special interest groups, such as cottagers' associations.

Tests performed by water-test technicians may include tests for pH; nutrients, such as phosphates, nitrates, and ammonia; major cations (Na^+, K^+, Ca^{2+}, Mg^{2+}); major anions (Cl^-, F^-); dissolved oxygen; bacteria; and algal content.

In recent years, computerized lab equipment has been developed that is capable of detecting very small concentrations (parts per billion) of trace organic contaminants, such as PCBs, dioxins, some aromatic hydrocarbons, pesticides, and herbicides.

Water-test laboratory technicians are usually graduates of a two-year community college program. They have the satisfaction of using their knowledge of analytical chemistry on a daily basis, and knowing that the work they do is helping to improve the environment.

Exercises

Key Words

acidity

dilute

hydrogen ions

hydroxide ions

indicator

Review

1. List three indicators you used and explain a positive and negative test for each in acid and base.
2. Name one indicator that measures an acid qualitatively and one that measures an acid quantitatively.
3. What effect does dilution have on the pH of a solution? Explain what happens during a dilution to alter the pH.
4. Name three household products that are bases and three that are acids.
5. Are acids more corrosive than bases?
6. What ion is responsible for acidic properties? basic properties?
7. Read the label of a product that is strongly acidic. What antidotes are to be given if it is taken internally? Externally?
8. Read the label of a product that is strongly basic. What antidotes are to be given if it is taken internally? Externally?

Extension

1. Research the industrial production of sulfuric acid.
2. Explain the difference between a weak acid and a strong acid.

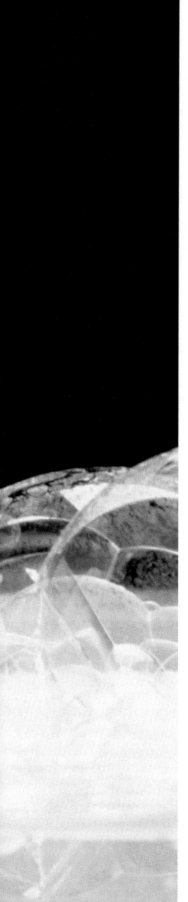

CHAPTER 14
APPLICATIONS OF
ACIDS AND BASES

You are probably wearing something that is made from an acid or from a product formed from an acid. For example, synthetic fibers, such as nylon, are manufactured using adipic acid. Many foods are preserved in acid.

As you walk to school in the rain or snow, you are being sprinkled with acid precipitation. The level of acids and bases in the environment is very important to living things.

In this chapter you will be experimenting with some important applications of acids and bases and learning about how they affect our lives and environment.

Key Ideas

- Acids are formed in the atmosphere when non-metallic oxides react with water.

- The excess acidity of precipitation has become a threat to our environment.

- Certain plants have a specific pH range at which they flourish.

- Soap is made from a strong base and fats.

- The food industry relies on acids for the taste and preservation of food.

- Neutralization reactions are being used to treat acidic lakes.

The soap that formed these bubbles was made using bases.

14.1
ACIDS AND BASES IN OUR ENVIRONMENT

For millions of years, acids have been made in the atmosphere. We are showered with them whenever it rains or snows.

The normal acidity of rain is caused by carbon dioxide dissolving in water in the atmosphere to produce a weak acid called carbonic acid, which is expressed as follows:

$$CO_2(g) + H_2O(l) \longrightarrow H_2CO_3(l)$$

Blow through a straw into water containing bromthymol blue. What color does the solution become? What gas has been exhaled?

Today the acidity of rain has become a real threat to our environment. It is caused by many gaseous pollutants, such as automobile exhaust fumes and the emissions produced by industry.

Table 14.1   Sources of acid rain in our environment

Source	Oxide Present	Acid Produced
car exhaust, lightning	oxides of nitrogen	nitric acid
coal, smelters, volcanoes	oxides of sulfur	sulfuric acid

Bases are not as great an environmental threat as are acids. Since bases are antacids, they have the ability to combat the acids in our environment. Bases are commonly used to neutralize or remove acids.

When lime, a base with the chemical name calcium oxide, is dumped into a lake affected by acid rain, the acid properties are neutralized.

14.2 ACTIVITY
HOW ACIDIC IS YOUR PRECIPITATION?

Important information can be gathered by collecting rainwater regularly from sites across the country. Normal rain is slightly acidic (pH about 5.6) because of the formation of carbonic acid from carbon dioxide in the atmosphere. Acidic precipitation is caused by both natural and artificial sources. When the two sources are combined they can have serious effects on the environment. In this activity you will monitor the precipitation in your area to determine if there are any trends and variations in the levels of acid precipitation. Your job is to decide whether there is a certain time of the year that is worse for acidic precipitation and if so, why.

Filter your samples to remove any suspended particles.

Materials

glass or plastic juice containers (6)
distilled water
filter paper and funnel
universal indicator paper
pH indicator solutions
test tubes (4)
glass stirring rod

Method

1. Clean the containers thoroughly and rinse them with distilled water. Place a container outside to collect rain or snow.
2. Pour your sample through a filter paper and funnel, and collect your filtrate in another container.
3. Using your stirring rod extract a sample of your precipitation and touch it to a piece of universal indicator paper. Record the pH of the sample.
4. Take your sample and divide it into four equal amounts in four test tubes.
5. Check with your teacher as to which pH indicator solution you should use. Test your four solutions with indicators that show color changes in the pH region of your sample. Add two drops of each indicator to your samples.
6. Check the pH of new samples over a few months and determine whether there is any relationship between the acidity of the precipitation and the time of the year.

Questions and Conclusions

1. What are the sources of the gases that cause acid rain?
2. Was there any relationship between acidity and time of the year?
3. Use your library resource center to find out the effect of acid rain on soil, water, fish, and plants.
4. How could you as an informed citizen reduce the effects of acid rain?

SCIENCE & SOCIETY

Factory smokestacks are one source of acid rain.

Acid Precipitation

Sulfur dioxide is a colorless, poisonous gas that dissolves in water to produce an acidic solution. Sulfur dioxide is formed when sulfur-containing coal is burned or when sulfide ores are smelted. It is also released by volcanic eruptions. When it burns, sulfur combines with the oxygen in the air to form sulfur dioxide. The equation for this reaction is

$$S(s) + O_2(g) \longrightarrow SO_2(g)$$

If the sulfur dioxide comes into contact with water or water vapor, it forms sulfurous acid:

$$SO_2(g) + H_2O(l) \longrightarrow H_2SO_3(aq)$$

Not all sulfur forms sulfur dioxide when it is burned. Under certain conditions it will form sulfur trioxide, SO_3.

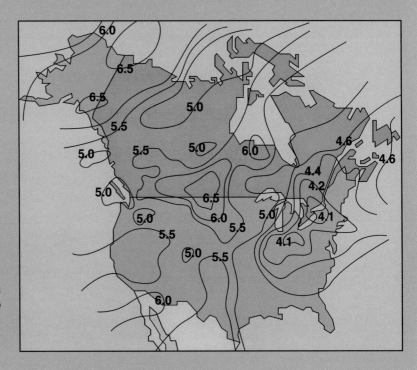

Figure 14.1
The numbers on this map indicate the pH of precipitation in various areas of North America. The areas with the lowest pH tend to be downwind of heavy industrial areas and large cities.

Sulfur trioxide unites with water in the air to form the extremely corrosive acid sulfuric acid:

$$SO_3(g) + H_2O(l) \longrightarrow H_2SO_4(aq)$$

The acids formed when these oxides combine with water vapor in the air eventually fall to the ground as acid rain or **acid snow**.

Acid rain and acid snow can raise the normal acidity of the soil and water to levels that are unsuitable for plant and animal life. They also can have corrosive effects on many types of building materials and artwork, such as statues and monuments.

Oxides of nitrogen, such as nitrogen monoxide **(NO)** and nitrogen dioxide **(NO_2)**, are formed by lightning and as combustion products of car engines. Nitrogen dioxide reacts with water vapor in the air to form nitric acid:

$$NO_2(g) + H_2O(l) \longrightarrow H_2NO_3(aq)$$

Oxides of sulfur and nitrogen can be carried a long way through the atmosphere by prevailing winds. A local pollution problem can readily become a widespread acid precipitation problem. Pollution produced in one country can even result in acid precipitation in another country.

These limestone sculptures were damaged by sulfur dioxide, a primary component of acid rain.

14.3
THE pH OF SOIL

Soil is the product of decomposition of organic material by bacteria and fungi, heat, and weathering. Soil is composed of air, water, minerals, and organic material. The balance of each of these components determines the pH of soil. The pH is important because it determines the availability of nutrients to plants.

When soil is too acidic, metal ions present in soil, such as magnesium, react and are carried or leached away, making them unavailable to plants. If the other extreme occurs and there is too little acid present, insoluble hydroxides and carbonates form, again making the metal ions unavailable.

The fish and plants in this pond have been killed by acid rain.

14.4   ACTIVITY
WHAT IS THE pH OF YOUR GARDEN?

Certain plants have specific pH ranges at which they flourish. Vegetables enjoy a slightly acidic soil whereas fruits enjoy a slightly basic soil. An acidic soil is often caused by the presence of active fungi while a basic soil is often due to the presence of minerals such as limestone. In this activity you will be testing the pH of garden soil and determining whether there is any correlation between the pH of a soil and the type of plant that grows best in the soil.

Materials

soil samples (4 x 25 g)
universal indicator paper
50 mL beakers (4)
boiled water (200 mL)
filter paper (4 Whatman #1)
microscope or hand lens

Method

1. Label the beakers 1, 2, 3, and 4.
2. Describe the physical and microscopic appearance of each of the four soil samples you are testing.
3. Place a piece of universal indicator across the bottom of each of the beakers.
4. Cut four pieces of filter paper so they will fit the bottoms of the beakers.
5. Place the pieces of filter paper on top of the pieces of universal indicator paper.
6. Add the four soil samples to the beakers.
7. Add enough cooled water to moisten the soil. (Do not soak the soil.) Allow the soil to sit for 3 min or until the water has moved through the soil to the bottom of the beakers.
8. Turn each cup upside down and dump the soil and universal indicator.
9. Record the color of the universal indicator paper and compare it to the color guide to determine the pH of the soil.

Questions and Conclusions

1. What were the pH values of your soil samples?
2. How did the pH of your soil samples relate to the physical appearance of the soil?
3. Was there any relationship between the type of vegetation grown and the pH of the soil?

14.5
SAPONIFICATION

The process by which a fat molecule is broken down into four smaller molecules when stirred with a strong base, such as sodium hydroxide, is called saponification. The reaction is illustrated below.

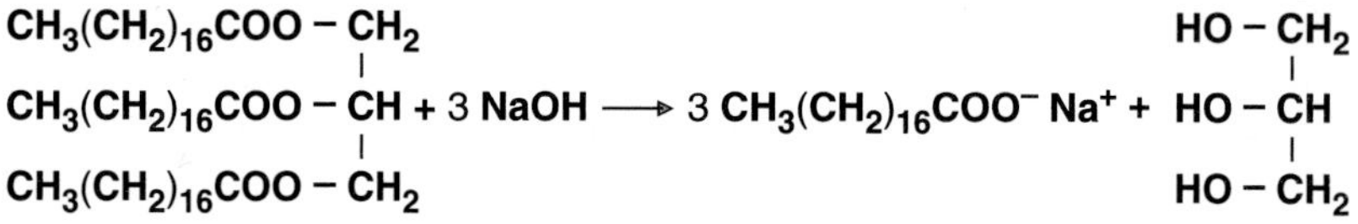

$$CH_3(CH_2)_{16}COO-CH_2$$
$$CH_3(CH_2)_{16}COO-CH + 3\ NaOH \longrightarrow 3\ CH_3(CH_2)_{16}COO^-\ Na^+ + HO-CH$$
$$CH_3(CH_2)_{16}COO-CH_2$$

| Stearic acid ester (Glyceryl tristearate) | Sodium stearate (a soap) | Glycerol |

HO – CH$_2$
HO – CH
HO – CH$_2$

Figure 14.2   Fat molecule and sodium hydroxide forming three soap molecules and a glycerol molecule

As chemists, we can understand the process behind soap formation. However, making a soap with the desirable consistency and fragrance is an art that is acquired after years of experience. The quality of the product depends on the temperature of the reactants, the stirring time, and the type of fat used.

14.6   ACTIVITY
PREPARATION
OF A SIMPLE SOAP

When you go into the cosmetics section of your local department store, there is an endless display of soaps. These soaps are finer quality soaps and have been prepared using high quality fats. In the early 1800s, the soaps used were harsh and were prepared by the same method you will use in this activity. The only difference was that in the 1800s, hydroxides were made by soaking ashes in water.

In this activity you will be preparing soap and comparing your results with those of your classmates.

Materials

10 g sample of fat	plastic spoon
sodium hydroxide solution	various scents
ethanol	paper cups
100 mL beakers (2)	pH meter or universal
thermometer	indicator paper
hot plate	

Caution: Sodium hydroxide is corrosive and causes burns. Handle it with care. Wear goggles and an apron. If you spill any on yourself, rinse it off immediately with large amounts of cold water and have your partner tell your teacher.

Ethanol is flammable. Keep it away from flames, sparks, and heat sources.

Do not use the soap you make on your skin.

School laboratory conditions do not meet pharmaceutical standards.

Method

1. Your teacher will give you a 10 g sample of a fat. Record the name of the fat for further reference.
2. Place the fat in a 100 mL beaker and heat it slowly until it melts.
3. Remove the beaker and allow it to cool to 40°C.
4. Mix the sodium hydroxide and ethanol in another beaker and heat it to 30°C.
5. Stir the sodium hydroxide/ethanol solution into the fat mixture. Stir constantly with a plastic spoon for approximately 10 min.
6. A thick substance will form. Add whatever scent you wish to your product.
7. Pour the mixture into paper cups and allow it to set for four days.
8. Test the quality of your soap by washing your glassware in the lab.
9. Use pH paper to test the alkalinity of your soap. (Alkaline is another term used for basic.)
10. Compare your results with those of your classmates.

Question and Conclusions

1. Which fat produced the soap with the best sudsing ability?
2. Did the pH of the product vary with the type of fat used?
3. Why is soap so effective for cleaning dirt?

14.7 ACTIVITY
THE CHEMISTRY
OF CABBAGE

Fall in many regions brings visions of changing leaves, and the harvesting of crops. When cabbage is plentiful it can be preserved as sauerkraut for an extended period of time. The sour taste of sauerkraut is caused by the presence of lactic acid, which is produced by bacteria fermenting the sugars present in the cabbage. We classify this as a chemical change.

Farmer's markets are a good source of fresh fruit and vegetables.

Materials

shredded cabbage	universal indicator paper
table salt	sterilized preserving jar

Method

1. Fill the preserving jar about half full of shredded cabbage.
2. Sprinkle 2 g of salt over the cabbage.
3. Fill the jar to the top with shredded cabbage.
4. Sprinkle another 2 g of salt over the cabbage.
5. Add boiling water to fill the jar and tap it to eliminate any air bubbles.
6. Measure the pH of the solution by removing some of the solution and testing it with universal indicator paper.
7. Cap the jar loosely and allow it to ferment for two to three weeks.
8. Test the pH of your final product.

Questions and Conclusions

1. Was there any change in the pH of the cabbage after two weeks?
2. At one point you tapped air bubbles from your jar. What was the reason for doing this?
3. What other foods might be made this way?

14.8 ESTERIFICATION

In the presence of certain concentrated acids, organic acids react with alcohol to form compounds called esters. Esters are compounds that have the following molecular structures:

$$R - O - C - R \quad (R = \text{remainder of the molecule})$$
$$\underset{O}{\overset{\|}{}}$$

Some of the aromas of fruits are due to the presence of esters. An ester usually has a pleasant smell such as that found in bananas or coconuts.

Table 14.2   Aromas, names, and formulas of some common esters

Aroma	Name of Ester	Formula
banana	isoamyl acetate	$CH_3 - C - C - C - O - C - CH_3$ with CH_3, H, H on top; H, H, H on bottom; O double bond
orange	octyl acetate	$C_8H_{17} - O - C - CH_3$ with $\overset{\|}{O}$
wintergreen	methyl salicylate	$CH_3 - O - C - C_6H_4OH$ with $\overset{O}{\overset{\|}{}}$

Try These

1. How is an ester made?
2. Why are esters desirable?
3. What types of acids are required to make an ester?

14.9 ACTIVITY
HOW SWEET IT IS!

In this activity you will prepare some simple esters.

Materials

hot plate	salicylic acid
tongs	amyl alcohol
test tubes (2)	concentrated acetic acid
400 mL beaker	concentrated sulfuric acid
methanol	

Caution: Your teacher will show you how to smell chemicals. Acids and bases are corrosive and cause burns. Handle them with care. Wear goggles and an apron. If you spill any on yourself, rinse it off immediately with large amounts of cold water and have your partner tell your teacher.

Methanol is flammable. Keep it away from flames, sparks, and heat sources.

Method

1. Place 2 mL of methanol in a test tube.
2. Add 1 g of salicylic acid.
3. Your teacher will add four drops of concentrated sulfuric acid for you.
4. Place the test tube in a beaker of water and bring the water to a boil.
5. Using the tongs to hold the test tube, smell the contents of the test tube as instructed by your teacher.
6. Add 2 mL of amyl alcohol to a second test tube.
7. Your teacher will add 2 mL of concentrated acetic acid and four drops of concentrated sulfuric acid to the test tube.
8. Place the test tube in a beaker of water and bring the water to a boil.
9. Smell the contents of the test tube as instructed by your teacher.

Chemicals must be smelled properly.

Questions and Conclusions

1. What odor did you detect in step 5?
2. What odor did you detect in step 9?
3. State three practical uses of esters.

14.10 ACTIVITY
ACIDS AND BASES IN ACTION

Acids are used to make nylon products.

Acids and bases surround you. You are probably sitting in a chair that at one point in its manufacturing was cleaned with an acid. Adipic acid probably was used at one stage in the manufacturing process of the clothes you are wearing.

Bases have made cleaning the kitchen a much easier task. Lime is sometimes used to line football fields and whitewash fences. You use a base every morning when you wash your hands with soap.

The following suggested topics represent common uses of acids and bases. With the help of the library resource center, research one of the topics and prepare a one-page report. Your teacher will supply you with more details.

Suggested Topics

1. Acids and bases in household cleaners
2. How the stomach digests proteins
3. Acids and the prevention of food spoilage
4. The use of acids in fertilizers
5. The use of nitric acid in the manufacture of explosives
6. The use of sulfuric acid in the refining of petroleum
7. The use of acids in batteries
8. The use of sulfuric acid in paints and pigments
9. The use of acids in swimming pools
10. The use of bases in the production of soap
11. Bases used in the textile industry
12. Bases used in the leather industry
13. The use of acids and bases in setting dyes in fabrics

OPPORTUNITY

Water-test Field Technician

Water-test field technicians, as the name implies, do most of their work outdoors. They are often involved in testing water samples from lakes, rivers, and streams in recreation areas. For this reason, they must be out on the water much of the time.

Field technicians play an important role in determining the condition of the water in these areas. It is important to know about the quality of the water because cottagers and residents in the area usually get their tap water directly from the nearest body of water.

The technicians perform a wide variety of routine chemical tests to determine the condition of the water. Tests include: pH tests; tests for nutrients, such as phosphates, nitrates, and ammonia; tests for major cations (Na^+, K^+, Ca^{2+}, Mg^{2+}); tests for anions (SO_4^{2-}, Cl^-, F^-); tests for dissolved oxygen; tests for algae content; tests for bacteria and plankton. Sometimes, field technicians are responsible for obtaining water samples that must be taken to a laboratory for more complex testing.

Field technicians are usually graduates of a two-year community college program. For anyone who enjoys being outdoors and spending their leisure time at outdoor sports such as fishing, this could be the ideal occupation.

Exercises

Key Words

acid snow

ester

saponification

soap

Review

1. Write the equation for the formation of acidic precipitation.
2. "Unpolluted rain is acidic." Explain this statement.
3. How do automobile exhaust fumes form acid rain?
4. What effect does pH have on the following?
   a. terrestrial ecosystems   b. aquatic ecosystems
5. Write a word equation for the formation of soap.
6. Define "saponification."
7. Explain the importance of acids in the
   a. food industry   b. perfume industry

Extension

1. Why are certain regions of North America more resistant to the effects of acid rain? Perform an experiment to show the effects of acid rain on various types of soil and rock.
2. Find out how to make kimchi, Korean pickled vegetables.

Petroleum Process Operator

Gasoline, kerosene fuel, lubricating oils, and petrochemicals are all produced by oil refineries. Some of the processes used to produce these products include distillation, extraction, cracking, reforming, and polymerization.

As a petroleum process operator, you would be expected to have a working knowledge of the above processes. You would be responsible for reading the processing schedule and laboratory recommendations. You would have to adjust or maintain the controls that coordinate the amounts of products or reactants in the process.

In order to control the operation, you would be reading and recording temperatures, pressures, and flow meters. You would be expected to recognize any abnormal readings. Your knowledge of solvents would be an asset in the maintenance of the interiors of the processing units.

Key Words

addition
condensation
cracking
hydrogenation
monomer
oxidation
polymerization
saturated fat
unsaturated fat

Exercises

Review

1. What is the common name for ethyne?
2. Define and give an example of each of the following reactions for hydrocarbons:
   a. cracking   b. hydrogenation   c. oxidation
3. What causes arteriosclerosis?
4. What is the link between cholesterol and arteriosclerosis?

Extensions

1. Use your library to research the invention of cellophane, rayon, nylon, or Dacron.

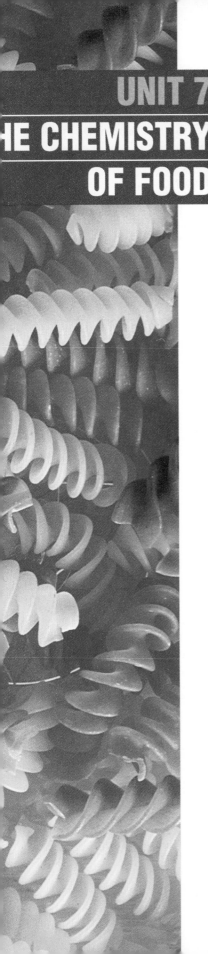

CHAPTER 17
WHAT'S
IN FOOD?

Biochemistry refers to the chemical reactions that take place in living organisms. Therefore, biochemists are interested in the kinds of molecules that make up living systems and the molecules from which they obtain their energy. Biochemists also are interested in the chemical reactions that occur in digestion, growth, maintenance, repair, and reproduction.

In this chapter you will consider some of the classes of compounds that are important to all living things, particularly humans.

Key Ideas

- All living things are made up of elements in the form of chemical compounds.
- Carbohydrates are the primary energy source for human beings.
- Lipids are very high in energy and have a variety of roles in the body.
- Proteins are made up of amino acids, which are needed by the body to make its own proteins.
- Foods can be tested for the types of nutrients they contain.
- Fiber is an important non-nutrient in our diets.

17.1 ACTIVITY
YOU ARE
WHAT YOU EAT

A wide variety of elements make up the human body.

All living organisms are composed of elements bonded together to form a wide variety of chemical compounds, some of which are very complex. Table 17.1 shows the various elements that make up and are required by the human body. Note that the first six of these elements make up 99% of the total.

Table 17.1   Elements found in the human body

Element	Chemical Symbol	Percentage
oxygen	O	62
carbon	C	20
hydrogen	H	10
nitrogen	N	3
calcium	Ca	2.5
phosphorus	P	1.14
chlorine	Cl	0.16
sulfur	S	0.14
potassium	K	0.11
sodium	Na	0.10
magnesium	Mg	0.07
iodine	I	0.01
iron	Fe	0.01
trace elements	—	0.8

Many of the elements listed are required in very small quantities by the human body, but they play very important roles. Other elements are required in greater amounts.

In this activity you will be examining labels for the chemical composition of various foods and matching the labels with the type of food.

Materials

list of foods examined
list of ingredients for each food

Method

1. Match the food in Column 1 with the ingredients in Column 2.

Column 1

taco shells

crackers

macaroni and cheese

gelatin dessert

marshmallows

pabulum

canned ravioli

chicken soup

Column 2

A: corn syrup, sugar, corn starch, water, gelatin, artificial flavor

B: ground corn, calcium hydroxide, hydrogenated soybean oil, salt

C: enriched wheat flour, modified milk ingredients, calcium chloride, lipase, salt, sodium phosphates, citric acid, tartrazine, lactic acid

D: sugar, gelatin, adipic acid, sodium citrate, fumaric acid, salt, flavor, color, tricalcium phosphate, maltol

E: oat flour, wheat flour, corn flour, rice flour, soya protein concentrate, dicalcium phosphate, barley, iron, niacinamide, thiamine (vitamin B_1), riboflavin (vitamin B_2)

F: wheat flour, skim milk, vegetable oil, salt, yeast, ammonium bicarbonate, baking soda, cream of tartar, butter, paprika, celery seed, spices

G: corn syrup solids, flour, starch, salt, vegetable oil shortening, monosodium glutamate, dehydrated cooked chicken, freeze dried chicken meat, chicken fat, seasoning, yeast extract, sugar, sodium caseinate, mono and diglycerides, spice, sodium citrate, dipotassium phosphate, carageenan, sodium stearoyl-2-lactylate, color

H: water, tomato puree, wheat flour, beef, toasted wheat crumbs, salt, carrots, cornstarch, sugar and/or dextrose, soya protein, vegetable oil, monosodium glutamate, caramel, seasoning and cheese flavor

Questions and Conclusions

1. List five unfamiliar materials in the ingredients list.
2. Using a chemical dictionary, look up what the five unfamiliar materials are used for.
3. Use a dictionary to define the term "food additive."
4. Name a food source that contains calcium.
5. Name a food source that contains sodium and chlorine.

Extension

1. Explain the difference between a food that is fortified and one that is enriched. Give an example of each.

17.2 NUTRIENTS AND WASTE

Food is a term used to describe materials that are ingested or eaten by living organisms. Food is classified as nutritious or non-nutritious.

Nutritious parts of foods consist of molecules that are required and can be digested by an organism. These parts are called nutrients. Non-nutritious parts of foods, such as the cellulose in celery, cannot be digested by most organisms. These parts are called wastes and are eliminated from the body by excretion.

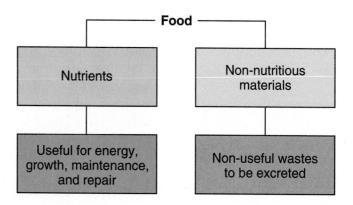

There are six nutrients required for a healthy human body: carbohydrates, lipids, proteins, water, minerals, and vitamins. In the following sections we will study the first three of these, while the rest will be discussed in Chapter 18.

17.3
CARBOHYDRATES

Carbohydrates are an extremely large class of chemical compounds made up of carbon, hydrogen, and oxygen. Sugars, starch, and cellulose are classified as carbohydrates. Two of these compounds, sugar and starch, are the primary energy sources for human beings. These carbohydrates are contained in foods such as corn, bread, potatoes, and fruit.

Carbohydrates can be found in these foods.

Carbohydrates always contain hydrogen and oxygen atoms in the same ratio as a water molecule. The name carbohydrate reflects the components of the compound. "Carbo" refers to the element carbon, and "hydrate" is from the Greek word for water.

Carbohydrates exist as monosaccharides or simple sugars, disaccharides, and polysaccharides. The most common monosaccharides are glucose, fructose, and galactose. They are the building blocks of all polysaccharides. The most common disaccharides are sucrose, maltose, and lactose. Polysaccharides are made up of many sugar molecules linked together by chemical bonds and are called starches.

Disaccharides and polysaccharides must be broken down into less complex sugars before they can be used by the human body. This chemical breakdown is part of the digestive process and is aided by digestive juices called enzymes present in the mouth and stomach.

The name for the chemical decomposition of carbohydrates is hydrolysis. Hydrolysis occurs, for example, when a disaccharide is mixed with water and is broken down by the water into two monosaccharides as shown:

$$C_{12}H_{22}O_{11} + H_2O \longrightarrow C_6H_{12}O_6 + C_6H_{12}O_6$$

disaccharide + water $\longrightarrow$ monosaccharide + monasaccharide

All disaccharides and monosaccharides contain carbon, hydrogen, and oxygen. The chemical equation on page 307 describes the hydrolysis of any disaccharide into any two monosaccharides.

However, specific disaccharides do produce specific monosaccharides, as follows:

Sucrose   +   Water

Glucose   +   Fructose

Glucose is the monosaccharide most frequently used by the human body as its energy source. Fructose is found in fruits, and galactose is the sugar found in milk.

Try These

1. What two components make up food?
2. Name the building blocks of starches.
3. What is the name of the process by which polysaccharides are split into simple sugars?
4. What controls the reactions from your answer to question 3?
5. What is the general formula for a monosaccharide or simple sugar?
6. What is the chemical formula for a common disaccharide?
7. What makes up a disaccharide?

SCIENCE & SOCIETY

Artificial Sweeteners

Artificial sweeteners are synthetic substances used in foods and beverages in place of sugar. They are lower in calories and sweeter than sugar. They are often used by people who are dieting and also by diabetics.

Saccharin, which is about 300 times as sweet as sugar, was first manufactured in 1879. It is made from petroleum products. Its use became widespread as a sugar substitute for diabetics.

Artificial sweeteners, like other foods and drugs, are carefully tested. Saccharin was tested both in Canada and the United States. Tests indicated that rats given extremely large doses of the sweetener developed cancer.

Other sweeteners, such as dulcin and P-4000, were banned because of similar results. Cyclamates, which also produced similar results as saccharin in tests in the United States and Britain, have been banned in some countries.

Aspartame was first produced in 1965. It is made from aspartic acid and phenylalanine, which are naturally occurring chemicals in foods. It is about 200 times as sweet as sugar. Aspartame is the most commonly used sweetener today. It is used in breakfast cereals, chewing gum, and carbonated beverages. It is considered to be safer than other artificial sweeteners but may cause problems for people who cannot tolerate phenylalanine.

Sugar Sweet

Low Calorie Sweetener
One packet equals
one teaspoon of sugar

Per packet: Energy, 1 calorie/4.184 kj;
Protein, 0.00035 g; Fat, 0 g;
Carbohydrate, 0.23 g; Aspartame 17.5 mg.
Ingredients: Dextrose, Aspartame
(contains phenylalanine).

The sugar substitute of choice

17.4
LIPIDS

Lipid is the chemical name for fats and oils. At room temperature, fats are solids and oils are liquids. Lipids have a higher ratio of hydrogen to oxygen than carbohydrates. The fat molecule glyceryl stearate ($C_{57}H_{110}O_6$) is an example.

Lipids are composed of a molecule of glycerol and three fatty acids. They are an even better source of energy than carbohydrates because they provide more than twice as much energy per gram. Like carbohydrates, they must be broken down into smaller units before they can be used for energy by the body.

If more fat is taken into the body than is needed, it is stored under the skin. Having a certain amount of body fat is important because it protects the body against the cold and acts as a cushion against hard blows. In North America, butter, cheese, vegetable oils, and meats are the main sources of fats in foods.

Try These

1. Which nutrients are the main sources of energy for the body?
2. What are the building blocks of lipids?
3. List some of the main sources of lipids.
4. Which type of nutrient provides the most energy per gram?
5. Why is body mass likely to increase if a person's main energy source is from lipids?
6. How does the ratio of hydrogen to oxygen compare in both carbohydrates and lipids?

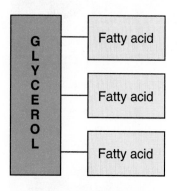

Bacon contains a lot of fat.

17.5
PROTEINS

Proteins, like carbohydrates and fats, contain carbon, hydrogen, and oxygen. Unlike carbohydrates and fats, all proteins contain nitrogen and most also contain sulfur.

Proteins are gigantic molecules made up of hundreds, sometimes thousands, of amino acids. They also must be broken into their amino acid building blocks before they can be used by the cells of the body.

Amino acids are used to produce, maintain, and repair the cells that make up the body. In addition, all of the genetic or hereditary material in the living cell is composed of amino acids.

There are 22 different amino acids found in the human body. Eight of these cannot be manufactured internally and must be obtained from foods. For this reason, these amino acids are called essential amino acids. The eight essential

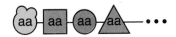

Figure 17.1   Amino acids (aa) are the building blocks of protein.

amino acids are listed in the following chart. A diet containing proteins provides the essential amino acids for a healthy body. Good dietary sources of proteins are lean meats, nuts, fish, milk, beans, cheese, and eggs.

Table 17.2   Essential amino acids

isoleucine	phenylalanine
leucine	threonine
lysine	tryptophan
methionine	valine

Try These

1. What is meant by the term "building blocks"?
2. What are the building blocks of proteins?
3. What is the role of protein in the body?
4. What is an essential amino acid?
5. How many amino acids are produced by the body?

These foods are good sources of protein.

17.6 ACTIVITY
TESTING
FOR FOOD
COMPONENTS

Different nutrients can be detected in foods using tests. For example, the presence of starch in a food can be shown by adding Lugol's iodine solution to the food. Lugol's solution has a characteristic brick-brown color that changes to blue-black when the iodine combines with the starch. This color change provides a positive indication that starch is present; that is why we call Lugol's solution an indicator.

The presence of lipids can be indicated by an optical effect. Fats and oils are large molecules that, like carbohydrates, have their atoms closely linked together. However, fat molecules are able to spread over the surface of an object. In addition, when spread thinly over a surface, fat molecules are translucent: they let light pass through. Therefore, translucence can be used as an indication of the presence of fat.

In this activity you will test foods for their content of starch, sugar, lipid, protein, and salt. Although salt is not a major component of food, it is essential in small amounts. Salt may also be added to enhance the taste of the food or as a preservative.

Part A

Starch Test

Materials

spot plate	pastry flour
eyedropper	saturated starch solution
whole wheat flour	unknown sample
enriched white flour	Lugol's iodine solution

Method

1. Put a small sample of each of the substances to be tested in a depression on the spot plate.
2. Put a drop of water into each of two depressions.
3. Add one drop of Lugol's iodine solution to each of the depressions containing a substance except for one of the two containing water. Put a drop of Lugol's iodine solution into an empty depression.
4. Record your observations. Compare what happens in the depressions containing water, iodine, water/iodine, and iodine/starch solution.
5. Test whether your unknown solution contains starch by putting a drop in one of the depressions and adding a drop of Lugol's iodine solution.

Part B

Sugar Test

Materials

large test tubes (6)	hot plate
eyedropper	safety glasses
test tube clamp	saturated sugar solution
test tube rack	sugar-containing foods (3)
thermometer	unknown sample
water bath at 37°C	Benedict's solution

Method

1. Set up a water bath and maintain it at 37°C.
2. Label four large test tubes. Put 5 mL of each test substance in the corresponding test tube. Put 5 mL of water in another test tube as a control.
3. Add ten drops of Benedict's solution to each of the tubes, and place each tube in the water bath.
4. Leave the tubes in the bath until no further observable changes take place.
5. Remove all the tubes from the bath and compare the test tubes containing the test substances with the tube filled with water and Benedict's solution. Record your observations.
6. Put 5 mL of your unknown solution in a test tube and add ten drops of Benedict's solution. Place the test tube in the water bath.
7. Repeat steps 4 and 5 for your unknown solution.

Part C

Lipid Test

Materials

porous paper	shortening
light source	vegetable oil
scissors	cream cheese
three types of nuts	unknown sample
butter	

Method

1. Cut a piece of absorbent paper into enough pieces so that there is one piece for each sample and one for water. Hold the absorbent paper at the edges only.

2. Label each piece of paper and pour 5 g of each test substance on its labeled paper. For dry ingredients, place a second piece of paper over the sample and apply pressure.

3. Let the samples stand for at least 15 min.

4. Carefully shake off or otherwise remove the material from the paper.

5. Compare the translucence of each sample with that of the vegetable oil paper by holding the paper up to a light source.

6. Test your unknown sample.

Part D

Protein Test

Materials

test tubes (4)	food sample solution
test tube rack	unknown sample
egg white	Biuret's solution

Method

1. Label four test tubes A, B, C, and D.

2. Pour 2 mL of raw egg white into tube A, 2 mL of your food sample into tube B, 2 mL of water into tube C and 2 mL of your unknown solution into tube D.

3. Add 5–10 drops of Biuret's solution to each of the test tubes.

4. Compare the color in each tube at the beginning and note any changes that occur within 5 min. Record your observations.

Part E

Salt Test

Materials

25 mL graduated cylinder
large test tubes (9)
test tube rack
filter
filter paper
goggles

table salt
foods containing salt
silver nitrate solution
unknown sample
distilled water

Caution: Silver nitrate is corrosive and causes burns. Handle it with care. Wear goggles and an apron. If you spill any on yourself, rinse it off immediately with large amounts of cold water and have your partner tell your teacher.

Method

1. In three large test tubes, prepare a 15 mL solution or suspension of each of the foods containing salt in distilled water. Shake vigorously.
2. Label three clean test tubes and pour each solution through a filter into a different test tube.
3. Prepare two test tubes, one containing a 15 mL solution of salt in distilled water and the other containing 15 mL of distilled water.
4. Add 5 mL of silver nitrate solution to each of the tubes. Mix each one thoroughly.
5. Observe each tube for the presence of a white precipitate, and compare with the distilled water sample. The white precipitate indicates the presence of chloride salts such as sodium chloride.
6. Put 10 mL of your unknown solution in a test tube and add 5 mL of distilled water. Shake vigorously and repeat steps 4 and 5.

Questions and Conclusions

1. Prepare a chart summarizing the nutrient tested for, indicator used, change observed, positive test, and negative test.

17.7 ACTIVITY
FIBER CONTENT

Cellulose, a polymer of glucose, is used by plants to strengthen their cell walls. Food coming from plants contains cellulose, but the human body possesses no enzymes capable of splitting this molecule, so we excrete it intact. As bulk, it is an extremely important component of our diet. Note that bulk, or **fiber**, is not considered a food because it cannot be digested. But fiber is extremely important to the operation of our intestines because it allows for easier and faster elimination of unwanted materials from our body. Some foods are better sources of fiber than others. Many breakfast cereals advertise "high fiber content." In this activity you will be testing these claims.

Materials

wash bottle	funnel and filter paper
balance	stretchable wrap or film
blender	10% amylase or diastase solution
beakers (4)	distilled water
glass stirring rod	samples of breakfast cereals (4)

Method

1. Put 25 g of one of the cereals and 50 mL of distilled water in the blender. Grind the mixture to a puree.
2. Find the mass of the filter paper, then put it in the funnel.
3. Add 50 mL of distilled water to the puree, stir, and filter.
4. Discard the filtrate and rinse the precipitate off the filter paper into a beaker using the wash bottle.
5. Add 50 mL of amylase or diastase solution. This will convert the starch into sugar, which is soluble.
6. Cover the beaker with film and store it in a warm place for 24 h.
7. Add 50 mL of distilled water and filter the mixture.
8. Discard the filtrate and allow the fiber to dry.
9. Measure the mass of the dry fiber and record the result.
10. Repeat steps 1–9 for the other samples.

Questions and Conclusions

1. Which breakfast cereal contained the most fiber?
2. Which breakfast cereals refer to fiber in their advertising?
3. Is there any chance of any other materials being present aside from fiber? How would this affect your results?

Exercises

Key Words

amino acid
carbohydrate
cellulose
disaccharide
energy source
fat
fiber
food
fructose
galactose
glucose
hydrolysis
lipid
monosaccharide
nutrient
nutritious
polysaccharide
protein
starch

Review

1. What is biochemistry?
2. What are the three most abundant elements in the human body?
3. Into what two categories is food separated by our digestive systems?
4. What happens to the nutritious parts of food inside the body?
5. What are the six main nutrients required by the human body?
6. What three kinds of molecules make up the carbohydrates?
7. Give an example of a disaccharide.
8. What are the building blocks of polysaccharides?
9. Name the process by which large organic compounds are split into smaller ones.
10. What two monosaccharides are produced by the hydrolysis of sucrose?
11. What are enzymes?
12. Which nutrients are classed as lipids?
13. What are the building blocks of lipids?
14. Which type of nutrient supplies the most energy per gram of the nutrient?
15. Why is a certain amount of body fat desirable?
16. What is the function of proteins in the body?
17. List three foods that are good sources of protein.
18. What are the building blocks of proteins?
19. What is an essential amino acid?
20. Match each of the following tests with the name of the nutrient for which it tests.

Test	Nutrient
paper test	starch
Biuret's solution	simple sugars
Benedict's solution	lipids
iodine solution	protein

21. In the starch test, what was the importance of testing the depression containing only water and the depression containing only Lugol's iodine solution?

CHAPTER 18
DIET
DETAILS

The human body is a very complex system. To work properly, it requires a wide variety of substances, some in very small quantities.

To get the energy and nutrients we need, we need to eat well-balanced diets. Too little of some nutrients can lead to various diseases, whereas too much food can also be unhealthy.

In this chapter you will look at some micronutrients (ones needed in small quantities) and the balance between energy intake and output.

Key Ideas

- Water is an essential part of all cells.

- Minerals and vitamins are essential in small amounts.

- Iron is often added to foods such as cereals.

- Our bodies need different amounts of energy for different activities.

- Whether you gain or lose mass depends on the balance between your energy intake and output.

- Large molecules, such as proteins, fats, and starches, are digested by enzymes.

18.1
OTHER IMPORTANT NUTRIENTS

Water

Water is an essential part of every cell in the human body. It dissolves almost any other soluble substance within the body, transports nutrients to all cells in the body, and is involved in most of the body's chemical reactions. This is why a person can live for quite a long time without food, but can survive only a few days without water.

Minerals or Inorganic Salts

Some **minerals,** or inorganic salts, are essential for maintaining normal bodily functions and growth. They are sodium, magnesium, phosphorus, chlorine, potassium, calcium, manganese, iron, cobalt, copper, zinc, and iodine.

Most of these elements are present in a balanced diet. The ones that are most likely to be deficient are iron, calcium, and iodine.

Table 18.1   Some essential minerals

Elements	Biological Effects	Dietary Source
calcium (Ca)	production of bones and teeth	asparagus, beans, cauliflower, cheese, cream, egg yolk, milk
iodine (I)	regulation of metabolism	broccoli, fish, iodized salt, oysters, shrimp
iron (Fe)	formation of hemoglobin	bran, cocoa, liver, mushrooms, oysters, peas, pecans, shrimp

Vitamins

Vitamins are substances that are required in small amounts by the body. They are not manufactured internally and must be obtained in the foods we eat or by means of vitamin supplements in the form of pills. However, if a person eats a well-balanced diet, vitamin supplements are usually unnecessary. Table 18.2 lists some important vitamins, the symptoms that signal their deficiency, and food sources from which they can be obtained.

Table 18.2   Some essential vitamins

Vitamin	Deficiency Symptoms	Important Sources
A (retinol)	inappropriate eye response to night light (night blindness)	green and yellow vegetables, fruit, dairy products, egg yolk
B_1 (thiamine)	weakness, paralysis	yeast, nuts, pork, whole grain cereals, liver
B_2 (riboflavin)	bloodshot, burning eyes	milk, cheese, eggs, yeasts, leafy vegetables, liver
B_6	convulsions, dermatitis	fresh meat, eggs, whole grains, fresh vegetables, liver
C (ascorbic acid)	anemia, sore gums, loss of teeth, swollen joints (scurvy)	citrus fruits, tomatoes
D (calciferol)	stunted growth, bones do not harden properly	egg yolk, milk
E (tocopherol)	shortened red blood cell life	meat, egg yolk
K	slow blood clotting, hemorrhage	green vegetables

Prolonged absence or shortages of vitamins in diet can cause certain vitamin-deficiency diseases. For example, scurvy is a disease caused by a deficiency of vitamin C that, in the past, often affected sailors on long voyages. This disease caused them to become fatigued, to develop pain in their joints, and to have bleeding gums and loose teeth. In the 1500s, British seaman Richard Hawkins observed that eating lemons and oranges prevented scurvy among his sailors.

Almost two hundred years later, British physician Dr. James Lind conducted an experiment that proved lime juice or lemon juice actually prevented scurvy. Thereafter, he recommended that all sailors in the Royal Navy be given regular rations of lime juice. This practice became widespread among British sailors and is where the nickname "limey" came from.

18.2 ACTIVITY
EXTRACTING IRON
FROM CEREAL

Vitamins and minerals are not required in the same quantities as carbohydrates, proteins, and lipids, but they are an essential part of our diet. Minerals are inorganic compounds that are essential to health in small amounts. Iron is a constituent of hemoglobin, and is necessary for the transport of oxygen by red blood cells.

In this activity you will be examining the iron content of fortified breakfast cereal. It is often added to cereal in the form of powdered iron.

Materials

magnetic stirrer and stir bar
resealable plastic bag
iron-fortified cereal
600 mL beaker

Method

1. Pour one cup of cereal into a resealable plastic bag and crush.
2. Empty the contents of the bag into a 600 mL beaker filled with water.
3. Place a magnetic stir bar into the solution and mix for 15 min.
4. Pour the solution into a large waste container and retrieve the magnetic stirrer.
5. Observe the stirrer. Record your observations.
6. Repeat for the other cereals.

Questions and Conclusions

1. What did you observe on the stir bar?
2. Which cereal had the most iron?

Extension

1. Repeat the experiment using massed amounts of different brands of cereal and test the claims on the label regarding iron content.
2. How is the iron powder used by the body? Design an experiment to test how iron powder is dissolved in stomach acid.

18.3
DAILY ENERGY REQUIREMENTS

We have seen that carbohydrates and fats contain the energy our bodies need to perform normal bodily functions. These nutrients also provide us with the energy we need for our daily work and other activities. But how much nutrient do we need? What happens if we take in more energy than our bodies can use? Or what happens if we use more energy than we take in?

It is possible to determine the energy content of foods. If you could calculate the number of kilojoules of food energy you use in a day, you could determine the number of kilojoules you should be taking in. The number of kilojoules required per person each day varies greatly and is affected by age, body mass, sex, and other factors such as activity level. The excess energy value taken in, but not used, is stored in the body as fat.

Skiing burns up more food energy than does homework.

Energy is measure in joules, but kilojoules are normally used to avoid large numbers. For example, the average male's daily energy requirement is approximately 13 000 000 J or 13 000 kJ and the average female's daily energy requirement is approximately 9 000 000 J or 9000 kJ. Expressing this energy requirement in kilojoules makes the numbers easier to work with.

In order to maintain a stable body mass, the number of kilojoules input per day must equal the number of kilojoules output per day. When this does not occur, there is an imbalance and mass will be gained or lost. In addition, a great deal of the day's kilojoules can often be obtained in an unbalanced form if we are not careful. As an example, bread, chips, and cola are

Table 18.3   Energy requirements for some activities

Activity	kJ/h	kJ/min
sitting	480	8.0
standing	510	8.5
walking vigorously	1 074	17.9
jogging	4 200	70.0
running vigorously	17 400	290.0
cycling vigorously	2 900	48.3
tennis	3 200	53.3
homework	510	8.5
sleeping	336	5.6
skiing, cross-country	6 600	110.0
skating	2 190	36.5
football	4 710	78.5
hockey	3 720	62.0
gymnastics	1 200	20.0
swimming (crawl)	7 530	125.5
swimming (breaststroke)	8 460	141.0
swimming (backstroke)	9 420	157.0
dancing	1 320	22.0
table tennis	1 680	28.0

very tasty and high in carbohydrates but provide little or no nutrients that are required by the body.

This is not the only problem. Even when a healthy diet is being followed, more kilojoules often can be taken in per day than are required. This causes the body to gain mass that may be undesirable.

The best way to lose body mass is to increase your level of activity. However, the form of exercise that you choose should interest you enough that you will continue to do it regularly. In this way, you can continue to take in enough kilojoules to satisfy your desire to eat while maintaining a steady body mass.

18.4 ACTIVITY
THE ENERGY
CONTENT
OF FOODS

Carbohydrates and fats are excellent sources of energy. Starch is excellent as a long-term energy source and sugar is excellent as a short-term energy source.

Marathon runners or skiers usually eat a large amount of starchy foods for two to three days prior to their event. This allows them to store up a large quantity of chemical potential energy that they can use up over a long period of time. Sprinters, on the other hand, often will eat a chocolate bar shortly before their race because sugar supplies a burst of energy over a short period of time.

Since heat is another form of energy, the energy content of various foods can be compared when equal masses are burned. In this activity you will measure and compare the energy content of various foods.

The chemical energy stored in food is converted to heat when it is burned. In this activity, the heat from the burning food will raise the temperature of a given mass of water. The amount of heat gained by the water can be calculated by using the following expression:

$Q = m \times \Delta T \times c$

Q = heat energy
m = mass of water
ΔT = change in temperature
c = specific heat capacity of water

The specific heat capacity of a substance is the quantity of heat required to raise the temperature of one gram of the substance by one degree Celsius. The specific heat capacity for water is 4.2 kJ/kg • K or 0.0042 kJ/g • K. You will use the latter value since the mass will be measured in grams.

Materials

retort stand	string
burner	porcelain dish
wooden splint	sugar
small and large can	peanuts
clamp	Brazil nuts
thermometer	suet
balance	

Method

1. Record your observations and show the results of your calculations in a chart similar to the one shown below.

Type of Food _____	Quantity	Units
a. Mass of porcelain dish		g
b. Initial mass of porcelain dish plus food		g
c. Final mass of porcelain dish plus food		g
d. Mass of food burned [b - c]		g
e. Mass of water (1 mL = 1 g)	100	g
f. Initial temperature of water		°C
g. Final temperature of water		°C
h. Temperature change [g - f]		°C
i. Quantity of heat produced $Q = (m\Delta Tc) = [(d)(h)c]$		kJ
j. Quantity of heat per gram of food [i ÷ d]		kJ

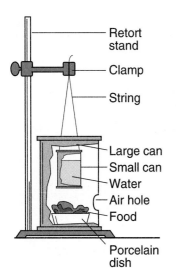

Retort stand
Clamp
String
Large can
Small can
Water
Air hole
Food
Porcelain dish

2. Determine the mass of the porcelain dish.
3. Place a piece or a few pieces of one type of food in the porcelain dish and determine the mass of the dish and the food.
4. Place 100 mL of water in a small can.
5. Set up the apparatus as shown in the margin.
6. Suspend the thermometer, using a string, so that it is in the water but not touching the sides or bottom of the can.
7. Record the temperature of the water.
8. Use a wooden splint to ignite the sample of food in the porcelain dish.
9. Once the food is burning, remove the splint.
10. Watch the thermometer and record the highest temperature reached by the water.

11. Remove the porcelain dish and its contents when the food stops burning and determine their mass once again.
12. Repeat steps 2–11 using different foods.

Questions and Conclusions

1. Which food tested contained the most energy per kilogram?
2. Which food gave off the least energy per kilogram?
3. Do you think the amounts of energy obtained in this experiment are accurate? Why?
4. List several possible sources of error in this experiment.
5. How would you improve the design of this experiment to make it more accurate?

18.5 ACTIVITY
PLAN
YOUR OWN DIET

In this activity you will measure your energy input and output over a three-day period, using the information given. Then, based on your results, you will prepare a personal diet.

Method

1. Copy the following chart into your notes.

| Day | Input | | Output | | |
	Type and Amount of Food	Energy Input	Activity	Duration of Activity	Energy Output
1					
2					
3					
4					
Total					

2. Using Tables 18.3 and 18.4, record your energy input and output for three consecutive days.
3. For each day that you record energy input and output, determine which is greater. Determine whether input or output is greater overall.
4. Decide whether you want to maintain your present mass, lose mass, or gain mass. Then plan a sensible diet of three meals a day to achieve this goal. The energy values for many types of food are listed in Table 18.4. If you eat something that is not in this table, estimate its energy content.

Questions and Conclusions

1. Which was greater, your total energy input or your total energy output?
2. Were your eating and activity habits during these three days typical? Do you usually eat more than this or less? Are you usually more or less active than this?
3. What would be the long-term effects of having an input greater than your output?
4. What would be the long-term effects of having an output greater than your input?
5. What should you do to make your energy intake equal to your energy output?
6. If your input is greater than your output, state two ways that you could change this.

Table 18.4   Food energy values in kilojoules

Fruits and Juices		Meat	
apple	330	bacon, 1 slice	210
apple juice, 250 mL	525	beef, roast, 100 g	1025
banana, medium	460	bologna, 1 slice	190
cantaloupe, half	250	chicken, drumstick	375
cherries, 125 mL	220	ground beef, 100 g	1025
fruit pie, 1 piece	1700	ham, 100 g	1045
grapefruit, half	190	lamb chop, medium	1670
grapes, 250 mL	270	liver, 100 g	1025
orange	300	pork sausage, 1	295
orange juice, 250 mL	500	pork spareribs, 3	290
peach	165	wiener	710
pear	475		
pineapple, 250 mL	310	**Breads**	
pineapple juice, 250 mL	565	bagel	630
plums, small, 2	165	bran muffin	405
raisins, 250 mL	2000	doughnut	590
tomato, medium	150	rye bread, 2 slices	690
		soda crackers, 3	175
Vegetables		waffle	270
carrots, raw, 250 mL	190	white bread, 2 slices	770
corn, ear	290	whole wheat bread,	
peas, 250 mL	480	2 slices	675
potato, medium	375		
potato chips, 50 g	1200	**Cereals and Pasta**	
		corn flakes, 250 mL	335
Milk Products		pasta, 250 mL	650
butter, 15 mL	420		
cheese, 1 slice	495	**Miscellaneous**	
cottage cheese, 250 mL	710	chocolate bar	1450
ice cream, 125 mL	690	egg	375
margarine, 15 mL	420	French fries, 20	1150
skim milk, 250 mL	425	pizza, 2 pieces	2600
2% milk, 250 mL	580	soft drink, regular	570
whole milk, 250 mL	750	sugar, white, 15 mL	190

18.6 ACTIVITY
ENZYMES OF DIGESTION AND THE HYDROLYSIS OF STARCH

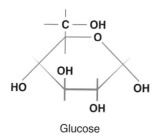

Glucose

Proteins, fats, and starches are all very large molecules. Before they can be used by the body they must be broken down into smaller fragments by a chemical process called **digestion**. Many different enzymes are involved in this process. **Enzymes** are molecules that regulate most of the chemical reactions that take place in living things.

An enzyme in human saliva called **amylase** begins the digestion or breakdown of the starchy foods we eat. In this activity you will examine the ability of this enzyme to hydrolyze starch molecules.

Starch is a very large molecule consisting of many glucose units joined together. As it is hydrolyzed, it is broken down into smaller and smaller fragments which eventually yield its building blocks, glucose molecules.

Materials

burner	thermometer
retort stand	400 mL beakers (2)
ring clamp	medicine droppers (2)
wire gauze	diastase solution
hot plate	starch solution (dilute)
test tubes (4)	iodine solution
test tube rack	Benedict's solution

Method

1. Design a data table to record your observations. Number the test tubes 1–4.
2. Set up a water bath on the hot plate and heat it to 37°C.
3. Set up a boiling water bath using the burner, retort stand, ring clamp, and wire gauze.
4. Put 10 mL of dilute starch solution into each test tube.
5. Put two drops of iodine solution into test tubes 1 and 2.
6. Put five drops of Benedict's solution into test tubes 3 and 4.
7. Put two drops of diastase solution into test tubes 2 and 4. (Test tubes 1 and 3 are your controls.)
8. Place all four test tubes into the 37°C water bath for 10 min. Note any color changes that occur.
9. After 10 min, place test tubes 3 and 4 into the boiling water bath for 5 min. Note any color changes that occur.

Summary of test tube contents				
Test Tube	Starch Solution	Diastase	Benedict's Solution	Iodine Solution
1	10 mL	none	none	2 drops
2	10 mL	2 drops	none	2 drops
3	10 mL	none	5 drops	none
4	10 mL	2 drops	5 drops	none

Questions and Conclusions

1. Explain the meaning of the color changes that took place.
2. What substance has been changed in test tube 2?
3. What substance has been produced in test tube 4?
4. Compare the contents of test tubes 2 and 4 and account for the changes that have taken place.
5. Explain the importance of test tubes 1 and 3.

Exercises

Key Words

amylase
digestion
energy value
enzyme
kilojoule
mineral
specific heat capacity
sugar
vitamin

Review

1. How many kilojoules of heat are required to raise 1 g of water from 20°C to 99°C?
2. Compare the amount of energy provided by 10 g of starch with the energy provided by 10 g of fat.
3. Compare the energy input and the energy output of a person who is slowly gaining mass.
4. a. Name the enzyme present in human saliva.
   b. What type of nutrient is split by this enzyme?
5. List any three essential elements for the body and one food that is a good source of each.
6. List any three elements and their biological effects on the body.
7. What is a vitamin? Give three examples.

Extension

1. Give reasons why a person might be overweight.
2. Why is it important to have protein in your diet?
3. Why is an exercise program important for someone trying to lose mass?

CAREER OPPORTUNITY

Diet Technician

One of the many interesting careers in the food industry is that of the diet technician. Diet technicians work in hospitals, restaurants, schools, and other public and private institutions.

For many hospital patients, an important part of coping with or recovering from illness is having a carefully planned diet. For healthy individuals, a well-balanced diet is essential for maintaining good health.

Diet technicians, working with dietitians, may use their knowledge of biochemistry and nutrition to plan menus and diets for hospital patients and residents in institutions.

Diet technicians work with the kitchen staff to purchase food and food-processing equipment, to maintain food cost control records, assist with inventories of supplies and equipment, and ensure sanitary conditions in the cooking and food storage areas.

Diet technicians must complete a two-year program at a college of agricultural technology or community college offering courses in food science.

UNIT 8
CONSUMER
CHEMISTRY

CHAPTER 19
THE CHEMISTRY
OF CLEANSERS

In earlier chapters you classified substances and chemical reactions, and analyzed products. Analysis and classification are necessary to ensure that consumer products are of high quality and are safe to use. Consumers need to be well informed in order to use consumer products effectively.

In this chapter you will use many familiar ideas to make, test, and classify consumer products.

Key Ideas

• Chemistry plays an important role in the manufacture of consumer products.

• Raw materials for consumer products are made available by chemical or physical means.

• The quality of consumer products is controlled by both industry and government.

• Many careers result from, and require an understanding of, applied chemistry.

• Evaluation of consumer products can be done using a scientific approach.

• Consumer products can be produced by the types of chemical reactions you studied earlier in this textbook.

19.1
CLEANSERS

"Wash your hands" is a saying we all grow up hearing. There is a valid reason for this saying: dirt carries germs that can cause disease and death.

Today, effective cleansers have been developed to eliminate disease-causing germs. Think of the many types of cleansers that you use every day. Soap is one, and there are dozens of different brands on the market. What makes them different? How do soaps differ from shampoos?

There are also many different types of detergents available for washing clothes and dishes. Some are biodegradable (capable of being broken down into harmless products by living organisms), others are not. What makes detergents different from soaps? What makes one type of detergent different from another? Aside from soaps and detergents, there are disinfectants, bleaches, and a number of different abrasive cleansers. These products can be classified into groups and the properties of the groups can be studied.

Try These

1. How can diseases be transmitted by dirty hands?
2. List four types of cleansers.
3. Use the library resource center to find two types of disease that are spread by lack of cleanliness.

19.2
TYPES
OF CLEANSERS

Each type of cleanser is unique and has its own properties that are worth studying and comparing.

Soaps

As you learned in Chapter 14, soaps are made by a process called saponification. Saponification occurs when a fat reacts with a base to form a soap and glycerin as follows:

$$C_3H_5(C_{17}H_{35}COO)_3 \quad + \quad 3\,NaOH \longrightarrow$$

fat (stearin)   base (sodium hydroxide)

$$3\,C_{17}H_{35}COONa \quad + \quad C_3H_5(OH)_3$$

soap (sodium stearate)   glycerine

Soaps cleanse because they have a special structure that can dissolve dirt particles. Each soap molecule has two oppositely charged ends. One end is soluble in water and the other end is soluble in oil. In this way, the soap acts as a bridge between the dirt or grease particle and the water molecule.

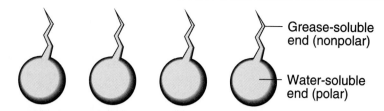

Grease-soluble end (nonpolar)

Water-soluble end (polar)

Figure 19.1   Soap molecules

Different soaps are made by combining various fats, oils, and bases. Some common soap salts are sodium stearate, sodium palmitate, and sodium oleate. Perfumes are often added to give the soap a pleasant fragrance.

When soap is used in hard water, a scum forms. This scum is caused by calcium ions present in the water. These ions react with the soap to form an insoluble salt precipitate.

An example of this type of reaction is as follows:

$$\text{sodium stearate} + \text{calcium chloride} \longrightarrow \text{calcium stearate} + \text{sodium chloride}$$

Calcium stearate is insoluble and clings to fabric as a solid. It also forms the insoluble scum that we see as a ring around our bathtubs. The undesirability of scum formation created a need to find a cleanser that would not react this way in hard water. Chemists solved this problem by inventing detergents.

Detergents

Scums or precipitates will not form if a detergent is used in hard water because the calcium and magnesium salts of detergents are soluble.

Detergents are made using an alcohol instead of a fat molecule and their cleansing action is similar to that of soaps. However, they outsell soaps by as much as three to one. In fact, they have almost replaced soaps, and are used in many shampoos and even in toothpastes.

The first detergents, however, were nonbiodegradable because they could not be broken down by the bacteria in the settling tanks of sewage treatment plants. This situation left the detergent chemically unchanged when it was released. The result was the collection of large amounts of foam in streams and ponds below sewage treatment plants.

The new technology had solved one problem and created another, as new technologies sometimes do. However, shortly thereafter, chemists developed a detergent that had a different structure that could be broken down by bacteria; that is, a biodegradable detergent.

Bleaches

A bleach is an unstable substance that decomposes easily and is used to change the condition of another substance. This usually means that the substance's color is changed.

Bleaches are generally used to remove dirt or grease from fabrics. They do this by oxidizing the color of the dirt or grease spot to a colorless substance. Sodium hypochlorite (NaClO) is a common active ingredient or oxidizing agent in bleach. It releases active oxygen that combines with a stain or dye to remove its color.

$$\underset{\text{colored}}{\text{NaClO(aq)} + \text{DYE}} \xrightarrow{\text{oxidation}} \underset{\text{colorless}}{\text{DYE [O]} + \text{NaCl(aq)}}$$

Disinfectants, Antiseptics, and Antibiotics

Disinfectants kill bacteria on contact. They are used to sterilize equipment and areas that must remain free of disease-causing bacteria. Lye and formaldehyde are disinfectants.

Antiseptics are weaker than disinfectants. They slow or stop the growth of bacteria, but do not kill bacteria on contact as do disinfectants.

Antibiotics are anti-bacterial chemicals produced by living organisms. They act like antiseptics because they slow or prevent the growth of bacteria.

Try These

1. Write a word equation describing how soap is made.
2. What property of a soap is useful in cleaning?
3. What makes one commercial soap different from another?
4. What disadvantage does soap have?
5. How are detergents different from soaps?
6. What is the main advantage that detergents have over soaps?
7. What was a disadvantage of early detergents?
8. How do bleaches work?
9. What are the differences between disinfectants, antibiotics, and antiseptics?
10. Compare the properties of
    a. soaps and detergents   b. bleaches and disinfectants

19.3 ACTIVITY
MARKETING
A NEW PRODUCT

In this activity you will look at the hypothetical reports of two companies as well as a bar graph. The bar graph summarizes the effects of each company's actions on the sales of their respective products.

Company A: Summary Report

1. Before year zero, five years were spent developing a detergent.
2. Research was carried out to develop a biodegradable detergent between years zero and four.
3. An extensive advertising campaign was carried out in years zero, two, seven, and twelve.
4. New management and a change in marketing strategy began in year nine.

Company B: Summary Report

1. Reaction to the development of a new product by Company A. Sale of a detergent similar to Company A's began in year one.
2. Research was carried out to develop a biodegradable detergent between years one and six.
3. An extensive advertising campaign was carried out in years one, three, five, and eight.
4. There was no change in management but a change in marketing strategy began in year 12.

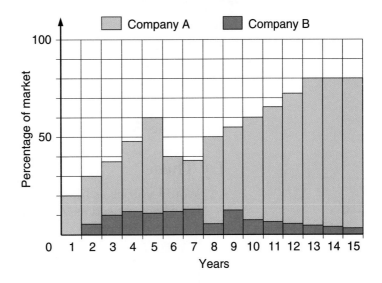

Method

1. Examine the summary reports and the bar graph for each of the companies over the 15-year period and answer the following questions.

Questions and Conclusions

1. Why is Company B one year behind in its introduction of a new product?
2. What effect might consumer pollution awareness have had on both companies' sales? Why?
3. How long did it take to change 50% of the people's minds about the advantages of biodegradable detergents?
4. Did the change in management have an effect on the sales of Company A's detergent?
5. What does your answer to question 4 tell you about the new management?
6. Between years eight and nine, which company had the largest percentage increase in sales? If you were in charge of promoting this company's product, what type of advertising claim could you make?

19.4

MARKETING METHODS MATTER

In Activity 19.3 you saw that throughout the 15-year period, Company B competed with Company A and kept a certain share of the market. This means that Company B manufactured a competitive product. The graph showed that a certain percentage of the public was convinced that Company B produced the superior product.

A new product must be marketed so that it will reach the majority of its intended consumers. If it is not, it will not sell nearly as well as a product that a great number of people have heard about. When was the last time you bought a chocolate bar or snack food for which you had not heard advertising? Just the mention of the words "taste test" brings the names of competing companies to mind. Advertising plays a large role in our choice of products.

Try These

1. Would you be as likely to pay to see a movie you had not heard advertised as you would a movie you had heard advertised? Why?
2. a. Would a magazine containing other consumers' opinions be of value to you in choosing a product? Why?
   b. Are there dangers in trusting other consumers' opinions about products?
3. What would be the role of a television program that tests consumer products and advertisers' claims?

19.5
SURFACTANTS—
SOAPS AND
DETERGENTS

Surfactants change the conditions at the surface site where one substance comes into contact with another. Detergents, soaps, and shampoos all form emulsions to remove dirt or grease that cannot be dissolved by water alone.

Recall that an emulsion is formed when two substances are made to dissolve that would not do so under normal circumstances. Milk, mayonnaise, and paint are examples. Emulsions are made up of colloids suspended in solution.

The nonpolar tail of a soap molecule is the oily hydrocarbon chain from the fat and is soluble in nonpolar substances such as dirt or grease. The polar head is the carboxyl group **(COO⁻)** and is soluble in water.

The nonpolar tails of the soap molecules line up and dissolve in the grease or dirt while the polar heads remain in the water. The soap works on the surface of the spot to break the grease or dirt into smaller particles and removes them from the material. These colloidal particles remain in solution because they are surrounded by a layer of polar heads that are attracted to water molecules.

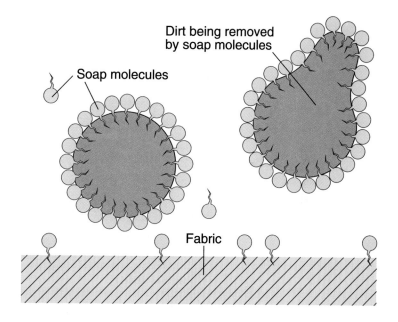

Figure 19.2   Soap at work

Try These

1. What is a surfactant?
2. What property of a surfactant is used in cleansing?
3. How does a soap form an emulsion?
4. Which end of the soap molecule is soluble in water?
5. How would the agitation of a washing machine help to lift grease or dirt from a fabric?

19.6 ACTIVITY
COMPARING SOAPS AND DETERGENTS

The main difference between soap and detergent molecules is the type of hydrocarbon tail that each has. To make detergents, a long chain alcohol is used instead of a fatty acid. The result is a cleanser that does not form insoluble compounds in hard water.

In this activity you will compare the action of a soap and a detergent.

Materials

50 mL beakers (2)	graduated cylinder
100 mL beakers (6)	1% commercial soap solution
large test tubes (5)	1% detergent solution
medicine dropper	distilled water
coins (2)	hard water
large stoppers (5)	mineral oil
pH paper or pH meter	cotton
ruler	flowers of sulfur

Method

1. Measure the pH of the soap and the detergent solutions. Record your results on a chart similar to the one below.

Property	Distilled Water	Soap	Detergent
pH	—		
distilled water: suds height	—		
hard water: suds height	—		
immediate emulsifying effect on oil			
emulsifying effect on oil after 5 min			
wetting ability			
surface tension			

2. Half-fill two large test tubes with distilled water.
3. Add 5 mL of soap solution to one test tube and 5 mL of detergent to the other.
4. Shake each test tube vigorously for 30 s. Measure and record the height of the suds formed. Rinse the test tubes.
5. Repeat steps 2–4 using hard water.

6. Add ten drops of mineral oil to each of the other test tubes.

7. Add 10 mL of soap solution to one test tube of mineral oil and shake. Repeat the procedure with 10 mL of detergent and then 10 mL of distilled water. Record your observations.

8. Observe each test tube from step 7 for 5 min and record your observations.

9. Place 50 mL of distilled water, soap, and detergent into separate 100 mL beakers. Add a piece of cotton to each. Compare the wetting ability of each.

10. Fill 50 mL beaker to the rim with distilled water. Note the shape of the meniscus (the curved surface of the water). Lower the edge of a coin into the water. Note the shape of the meniscus.

11. Add a few drops of soap solution to the water. Observe any change in the meniscus. Again lower a coin edgewise into the solution. Record your observations in the chart opposite "surface tension."

12. Repeat step 11 but use detergent. Record your results.

13. Place 50 mL of distilled water, soap solution, and detergent solution into separate 100 mL beakers. Sprinkle sulfur onto the surface of each. Compare the results.

Questions and Conclusions

1. Why might the pH of a cleansing agent be important?
2. a. How do soaps and detergents emulsify oil in water?
   b. How is emulsifying ability related to cleaning ability?
3. What effect do soaps and detergents have on surface tension?
4. How do soaps and detergent make water "wetter"?
5. What did the results of step 13 tell you?

19.7 ACTIVITY
CLEANING VALUE
FOR YOUR DOLLAR

Builders are added to detergents to soften water and to stop scum from forming. When scum forms on a fabric, it leaves a dull grey appearance. Phosphates are often used as builders but have two disadvantages. First, they act as buffers and keep the pH between 9 and 10. This can cause skin irritation and damage to fabrics. Second, they act as fertilizers and increase algal growth in lakes and ponds.

In this activity you will look at the cost of detergents from the point of view of cleansing efficiency. You will measure the amount of detergent needed to absorb a fat and use a standard flask for comparison.

In the standard flask, all of the fat is absorbed. Each detergent will then be compared to the standard flask to decide the endpoint. The less detergent needed, the better value it is, and the greater its cost efficiency.

Materials

hot plate
100 mL graduated cylinder
1 L Erlenmeyer flasks (7)
glass stirring rod
pH paper or pH meter
liquid laundry detergents (3)
lard colored with pink dye

Method

1. Make a 1% and a 3% solution of each detergent as instructed by your teacher. Using a pH meter or pH paper, measure the pH of the 1% solution of each brand and record it in a chart similar to the one shown below.

Brand	A	B	C
cost per mL			
pH (1% solution)			
dispersion volume			
volume of detergent			
cleansing cost			

2. a. Make the standard flask by heating 500 mL of water to 60°C in a 1 L Erlenmeyer flask on a hot plate.

   b. Add 1.0 g of colored lard to the water.

   c. Add 3 mL of liquid detergent to 22 mL of water in the graduated cylinder and stir.

   d. Add the detergent solution to the fat mixture in the Erlenmeyer flask. Keep the temperature at 60°C.

   e. Shake the flask vigorously until the lard is dispersed. You will compare each test flask to this standard flask.

3. Prepare a second flask by repeating steps 2a and 2b.

4. Add a 3% detergent solution 5 mL at a time to the flask and shake. Keep adding detergent solution until the fat is as dispersed as it is in the standard flask. This will give you the approximate volume of the detergent needed.

5. Repeat step 4 using smaller volumes of solution near the endpoint. Record the exact volume of solution needed. This is the dispersion volume.

6. Repeat steps 3–5 for each detergent to be tested.

7. Calculate the cost of the detergent in ¢/mL by dividing the cost by the volume of the bottle.

8. Calculate and record the actual volume of detergent used. The dispersion volume is 0.03 times this volume because the detergent used was 3%.

9. The cleansing cost is calculated by multiplying the volume of detergent used by the cost from step 7.

Questions and Conclusions

1. Which detergent would be best suited for fine fabrics or sensitive skin? Why?

2. a. Why are phosphates used in detergents?
   b. What are the disadvantages of phosphates?

3. Which detergent was found to be the most cost effective?

4. Do these results support any advertising claims you may have heard?

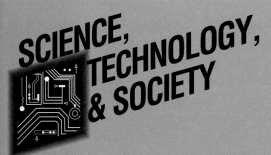

SCIENCE, TECHNOLOGY, & SOCIETY

Cleaning Up Oil Spills

In recent years, there have been some disastrous accidents involving oil supertankers. One of the worst occurred in March of 1989, when the *Exxon Valdez* ran aground off the coast of Alaska. Millions of liters of oil spilled into Prince William Sound. The spill seriously damaged a fragile ecosystem, which may take years to recover.

About six million tonnes of oil enter the world's oceans every year, adding to the 400 million tonnes already there. Half of this total comes from discarded used motor oil and street runoff. Most of the rest comes from small tankers. Some oil naturally enters the ocean from underwater crevices in the earth's crust. Thus, major oil spills like the *Exxon Valdez* are only the tip of the "marine oil iceberg."

What happens to oil once it enters the ocean? When crude oil is spilled into water, it forms a slick on the surface. Aided by the wind, the lighter fractions of oil evaporate into the atmosphere. Heavier fractions of the oil may sink. Sunlight acts to break down the oil. Wave action causes the oil to mix with the water. Eventually, bacteria break the oil down. In cold water or in sheltered coves, this process may take up to a decade.

Workers cleaning up the *Exxon Valdez* oil spill

Human action can speed up the natural clean-up processes. In calm water, the oil can be contained by floating booms and vacuumed from the surface. Oil can be deliberately set on fire to destroy it. Sometimes detergents are added to the oil to emulsify and disperse it.

Light oils allow water to enter the down layers of seabirds, reducing buoyancy and insulation. Heavy oil weighs birds down, making it hard for them to fly and swim. Sea otters depend on their dense fur to protect them from frigid water: light oils destroy the insulating properties of their fur. Shellfish living in coastal waters are vulnerable to smothering.

Marine plants can be killed when oil soaks into sediments or encrusts the plant. However, plants seem to recover more rapidly than animals, unless the water is chronically polluted.

All of this damage to marine organisms and plants can affect humans indirectly by reducing tourism and fishing. Contamination of shellfish and fish can lead to bans on commercial harvesting that can last for years.

However, marine ecosystems are capable of recovering from large oil spills. After the *Amoco Cadiz* disaster off France in 1978, it was predicted that the area would be lifeless for decades. Five years later, commercial fishing had resumed and it was hard to find obvious traces of oil. Unfortunately, oceans cannot absorb unlimited amounts of oil. We must search for ways to minimize the effects of oil on the environment.

19.8 ACTIVITY
BLEACHING

The active ingredient in household bleach is sodium hypochlorite **(NaClO)**. The hypochlorite ion **(ClO⁻)** is a good oxidizing agent. It oxidizes colored substances, such as stains or dyes, into compounds that are colorless.

This can easily be observed if pieces of wet, colored cotton are placed in bottles of chlorine gas. The wet cloth is quickly bleached because the water in the wet cloth reacts with the chlorine gas **(Cl₂)** to produce hypochlorous acid **(HClO)**, which is unstable. It oxidizes the dye in the colored cloth. The overall reaction is

$$H_2O(l) + Cl_2(g) \longrightarrow HClO(aq) + HCl(aq)$$

Aside from sodium hypochlorite, household bleach usually contains sodium hydroxide, chlorine, sodium chloride, and water.

Chlorine bleach is used to whiten fabrics and remove stains caused by ink, juices, and other substances. However, it cannot be used on all fabrics because its oxidizing action is too strong and can destroy delicate fibers such as silk and wool.

Bleach containers are labeled with the warning "Harmful gas formed when mixed with acid" or "Do not mix with acid." (What would happen if you did?) Bleaches also carry the corrosive warning symbol. Both the statements and the symbol indicate that you should be careful when using bleach.

In this activity you will test the action of household bleach on a variety of fibers and dyed fabrics.

Materials

evaporating dish
spot plate
medicine droppers
household bleach
variety of stains and dyes
variety of fabrics

Caution: Bleach is corrosive and causes burns. Handle it with care. Wear goggles and an apron. If you spill any on yourself, rinse it off immediately with large amounts of cold water and have your partner tell your teacher.

Method

1. Add ten drops of bleach to the depression of the spot plate for each stain that is to be tested.
2. Place a drop of stain or dye into the bleach in the depression of a spot plate. Record your observations in a chart similar to the one shown.

Stain or Dye	Color		Condition	
	Before	After	Before	After
A				
B				
C				
D				

3. Place a small piece of each fabric to be tested in an evaporating dish.
4. Add ten drops of bleach to each fabric. Observe the fabric for 10 min.
5. Rinse the fabric and record your observations.
6. Compare the condition of the fabric before and after bleaching.

Questions and Conclusions

1. What is the active ingredient in household bleach?
2. What type of reaction lets bleach remove stains?
3. a. Bleach cannot be used on all fabrics. Why?
   b. What fabrics are affected by bleach?
4. List the uses of bleach.

19.9 ACTIVITY PERCENTAGE CHLORINE AND THE COST OF DIFFERENT BLEACHES

Household bleach contains other substances in solution. Chlorine, sodium hydroxide, and sodium chloride are all present, along with water and sodium hypochlorite. There is an *equilibrium*, or balance, between these substances within the solution. When bleach is mixed with acid, the balance is destroyed because the acid reacts with the base, sodium hydroxide. This causes a shift in the equilibrium. As a result, chlorine, a poisonous green gas, is produced.

The above reaction is the first stage in the test for the percentage of chlorine available in household bleach. Then the chlorine produced reacts with potassium iodide according to the following equation:

$$Cl_2(g) + 2\ KI(aq) \longrightarrow 2\ KCl(aq) + I_2(s)$$

The number of moles of iodine produced will be the same as the number of moles of chlorine available. As iodine is formed, the solution turns brownish-yellow.

In this activity you will be comparing different brands of household bleach. Each group of students will be given one brand to test.

You will find the amount of iodine present by titrating the solution with sodium thiosulfate. When the brownish-yellow color has nearly disappeared, a starch solution is added to give it a blue color. This will give a clearer endpoint.

Materials

250 mL Erlenmeyer flasks (2)
graduated cylinder
burette
10 mL pipette with bulb
medicine dropper
household bleaches (3 brands)
potassium iodide
sulfuric acid solution
sodium thiosulfate solution
starch solution

Caution: Bleach and sulfuric acid are corrosive and cause burns. Handle them with care. Wear goggles and an apron. If you spill any on yourself, rinse it off immediately with large amounts of cold water and have your partner tell your teacher.
    Use the bulb and pipette to transfer the bleach. Do not use your mouth.

Method

1. Using a pipette, transfer 1.0 mL of household bleach to an Erlenmeyer flask. Use the bulb, not your mouth.
2. Add 75 mL of water to the flask.
3. Add 2 g of potassium iodide to the flask and dissolve it by swirling the flask.
4. Add 30 mL of sulfuric acid. Swirl the flask.
5. Fill a clean burette with sodium thiosulfate solution.
6. Record the initial reading on the burette. Titrate the bleach with the sodium thiosulfate solution.
7. When the solution turns pale yellow add ten drops of starch solution to the flask. The solution will turn blue.
8. Continue titrating until the blue color just disappears. Record the final reading on the burette.
9. Determine and record the volume of sodium thiosulfate added.
10. Repeat steps 1–9 for another sample of the same brand of bleach. If the volumes of sodium thiosulfate used in each titration are not within 1 mL of each other, repeat.

11. Average the volumes used in the titrations for the samples of your brand and record the results in a chart similar to the one shown below.

Brand	A	B	C
average volume			
percentage chlorine present			
cost (¢/mL)			
bleaching cost			

12. Record the class results for each of the three brands of bleach.
13. Use the graph provided by your teacher to determine the percentage of chlorine available in the bleach.
14. Calculate the bleaching cost by calculating the cost in ¢/mL divided by the percentage of chlorine present.

Questions and Conclusions

1. How did your experimental value compare with the value advertised on the label?
2. a. Which brand of bleach had the lowest bleaching cost?
   b. Are there other factors to consider when you determine the best buy in bleaches? What are they?
3. Why should bleaches and acids not be mixed?
4. a. Look for containers of bleach, acids, and ammonia in your home. Are they stored safely out of the reach of children?
   b. What precautions should be taken when you are storing bleach?

SCIENCE, TECHNOLOGY, & SOCIETY

Detergents and Phosphates

In chemistry, a detergent is used to lower the surface tension of water. Soap is one type of detergent and is made from animal fat or vegetable oil.

Manufacturers introduced synthetic detergents in the 1950s. These were not based on animal fat or vegetable oil but were based on petroleum products. The early detergents had branched structures and were found to be non-biodegradable. This had not been a problem with the earlier soaps because they had been made from animal and vegetable compounds which could be broken down by bacteria in rivers, lakes, and sewage treatment plants. As a result, detergent suds collected on the surface of rivers and lakes.

Applied chemists also found a way to stop scum from forming, making the detergent more effective. Additives were introduced to the detergent which would "soften" the water. As a solution to the cleansing problem in hard water, the applied chemists arrived at a reasonable solution. Science and technology moved from a natural but messy product to a high quality, effective, biodegradable, inexpensive, synthetic detergent. This sounds great but there was another problem.

The problem with phosphates in detergents is that they act as fertilizers for algae in the waters of the lakes and rivers that they enter. Algae growth increases rapidly. A bloom, or sudden growth, may cover the entire

Water pollution

water surface. The algal growth causes a depletion of oxygen in the water and results in the death of other forms of life in the water. Eutrophication of lakes occurs in a few decades rather than thousands of years. Phosphates in detergents became a major contributor to the phosphates accumulation found in lakes and rivers. Phosphates can only be removed in sewage treatment plants by the addition of expensive treatment.

Governments have limited the use of phosphates in detergents, even though new low phosphate or phosphate-free detergents can cost more to produce.

The algae on this lake may be a result of excessive nutrients.

One of the chemicals being used to replace phosphate is NTA, nitrilotriacetic acid. This new additive also has its hazards, which you may wish to research. It combines with elements like mercury and lead, keeping them in solution and thus in the food chain. What effects might this have?

OPPORTUNITY

Quality Control Technician

The role of a quality control technician is varied. Both qualitative and quantitative laboratory techniques are used to evaluate products. Materials are tested to make sure that the products meet the high quality standards of industry and government. The role of a quality control technician can include any of the following:

1. Collection and analysis of samples such as automobile exhaust, or exhaust from smoke stacks
2. Testing of water samples for pollutants
3. Product testing for purity and uniform quality

Quality control technicians are employed by companies that manufacture products for human consumption and personal use. Examples of these are the food and confection industries, the drug industry, and the cosmetics industry.

Exercises

Key Words

antibiotic

antiseptic

biodegradable

bleach

builder

cleanser

detergent

disinfectant

equilibrium

scum

surfactant

Review

1. Name four different kinds of cleansers.
2. Write the general equation for making soap.
3. Why does soap form scum with hard water?
4. Why do detergents not form a scum with hard water?
5. Distinguish between disinfectants, antiseptics, and antibiotics.
6. a. Define the term "emulsion."
   b. In what way does soap behave as an emulsion?
7. Explain the steps involved in cleaning a grease spot from a fabric.
8. Compare and contrast soaps and detergents.
9. How do soaps affect the following?
   a. surface tension
   b. wetting ability of water
10. What are the advantages and disadvantages of phosphates in detergents?
11. What is the active ingredient in household bleach?
12. How is oxidation important to the bleaching process?
13. Name three dyes or stains affected by bleach.
14. What steps are used to find the available chlorine in a bleach?

Extension

1. Write to a company that makes soaps and detergents and request information on its products, or visit a factory if there is one nearby.
2. Trace the development of detergents from early products to present products.
3. Imagine that you own a cottage at the other end of a lake from a soap manufacturing plant. You suspect that the plant is dumping phosphates into the lake. What could you do to find out? What other action would you take?
4. Are household acids, bases, and bleaches stored safely in your home? Report on the locations of each.

CHAPTER 20
PERSONAL
CHEMISTRY

In Chapter 19 you looked at some of the chemistry involved in very common consumer products—soaps and detergents. There is a wide variety of other consumer products that we use every day—products such as cosmetics, food additives, medicines, and other drugs.

In this chapter you will learn the chemistry that every informed consumer needs to know to make decisions about these products. You will also simulate the marketing of a product by designing advertising labels for it.

Key Ideas

- Consumer products can be classified as one of the types of matter discussed in Unit 1.

- Chemistry plays an important role in the manufacture of consumer products.

- Food and drug industries make use of chemical principles in the manufacture of their products.

20.1 ACTIVITY CLASSIFYING SOME CONSUMER PRODUCTS

In earlier chapters you classified and separated matter. Most consumer products are either homogeneous or heterogeneous mixtures. In this activity you will classify different types of cosmetics. Cosmetics are substances that are usually meant to improve the appearance of the person wearing them.

Cosmetics are mixtures of many types of materials.

Materials

foundation cream

lipstick

aftershave

rouge

bronzer

eyeliner

mascara

eye shadow

Method

1. Cosmetics can be thought of as one of five types of mixtures, as shown in the following chart.
2. Using the chart, classify each of the cosmetics as one or more of the five types listed. The same cosmetic may come in more than one form.
3. Use the definitions to state a reason for the category chosen.

Type of Mixture	Description
emulsion	a colloidal suspension of one immiscible liquid in another
powder	a mixture of minute solid particles with a talc base
wax/oil	a mixture of dyes and oils with a wax base
gel	a colloidal suspension that appears to be a solid
suspension	a dispersion of very small particles spread out evenly in a liquid

Questions and Conclusions

1. How does classifying matter help us?
2. What differences are there between heterogeneous and homogeneous mixtures?
3. Why should you read the label on a product?
4. How many classes of cosmetics have you studied?
5. What is the difference between an emulsion and a gel?
6. Name two reasons why cosmetics might be used other than to improve appearance.
7. The quality of a cosmetic, such as a foundation cream or a moisturizing cream, must be very high. Quality control technicians employed in this industry ensure that each product meets the standards set by the industry and the law. What are some of the factors that would influence the quality of a cosmetic applied to the face?

20.2 ACTIVITY PREPARING A FACE CREAM

In Activity 20.1 you classified some personal care products. As a rule, these products are meant to improve the appearance of the person using them. In this activity you will prepare a face cream in one of five ways and compare the results.

Cold creams, vanishing creams, and cleansing creams are usually emulsions. They contain oils, fats, waxes, and fatty acids in water. Lanolin from sheep's wool is most often used.

Materials

hot water bath
250 mL beaker
400 mL beaker
glass stirring rod
graduated cylinder
lanolin

mineral oil
stearic acid
triethanolamine
distilled water
perfume
preservative

Caution: Never heat cosmetics over a flame. Use a hot water bath.

Do not use the products you make. School laboratory conditions do not meet pharmaceutical standards.

Method

1. Follow the directions given for one of the five sets of ingredients. Your teacher will assign a set for each group.

Ingredients	Set 1	Set 2	Set 3	Set 4	Set 5
lanolin	9 mL	9 mL	9 mL	0 mL	9 mL
mineral oil	16 mL	13 mL	0 mL	13 mL	13 mL
stearic acid	13 mL	13 mL	13 mL	13 mL	0 mL
triethanolamine	3 mL	0 mL	3 mL	3 mL	3 mL
water	65 mL	65 mL	65 mL	65 mL	65 mL

2. Put the lanolin, mineral oil, and stearic acid into a 250 mL beaker and heat it in a water bath until the mixture melts.
3. Mix the triethanolamine and water in the 400 mL beaker and heat it in a water bath.
4. When the temperature has reached 85°C, pour the oil mixture from step 2 into the water a little at a time while stirring constantly. Stir until the paste is smooth.
5. Perfume your cream if desired, and add a general preservative as instructed by your teacher.

Questions and Conclusions

1. How does the texture and spreading ability of the creams you prepared compare to those of commercial creams?
2. Which set of ingredients do you think produced the best face cream? the worst? Why?
3. Which substance appears to be the emulsifying agent?
4. What is the purpose of a face cream?
5. What was the purpose of changing the amounts of ingredients in the five sets?

20.3
THE CHEMISTRY OF HAIR CARE

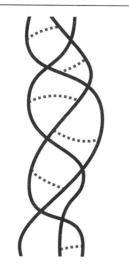

Figure 20.1
The structure of hair

Human hair is often thought to be a thing of beauty. North Americans spend billions of dollars a year on hair care and hair care products to keep their hair clean and attractive. Cleaning and styling hair involves many chemical principles.

Hair is made up of many protein chains intertwined like the strands in a rope. The chains are held in place by chemical bonds between the side branches of the chains.

Styling hair involves breaking some of the bonds between the side branches and forming new ones. If the protein chains do not form new bonds, the strands separate, causing split ends.

Wet hair is easier to style because water molecules help to break the bonds between the strands. As you comb hair, strands slide into new positions. As the water evaporates, the side branches form new bonds in the hair's new position.

Many hair stylists apply heat from hair dryers or curling irons. Heat speeds up the styling process because it speeds up the chemical reactions between the strands. Too much heat, though, can dry out hair and cause split ends.

20.4 ACTIVITY
DESIGNING LABELS

Labels list ingredients, give directions and safety precautions for a particular product, and also advertise the product. In this activity you will make your own label to advertise a personal care product of your choice.

Materials

personal care product labels
pen
craft paper
catalog

Method

Put yourself in the position of a marketing or advertising manager. Design a label to advertise your product. Here are some questions you might want to think about before you begin.

1. Is the product a new product or an improved one?
2. What does research say about the use of color in advertising?

3. What effects do the size of the label and the size of the container have on sales?
4. What age group is your target market?
5. What laws control advertising of the product?
6. What special claims can be made about the product?
7. In what way is the product unique?

Questions and Conclusions

1. How does knowledge of a product help the person advertising it?
2. Why is a knowledge of chemistry important when this type of product is advertised?
3. What role does color play in advertising?
4. What might the words "new" or "improved" mean on your label?
5. Do you think that labels should list all ingredients? Why or why not?
6. What effects do you think the use of a film or television celebrity in an advertisement would have on the sales of a product? Have you ever been influenced to buy a product that your favorite celebrity advertised?

20.5 FOOD ADDITIVES

Food has never been more safe nor contained more chemicals. Some chemicals preserve food and protect the people who consume it. Other chemicals are added that may be dangerous. The problem is that little is known about the effects of small doses of these substances over long periods of time.

A food additive is a substance or its byproduct, including radiation, that may become part of, or affect, the properties of a food. This does not include natural ingredients such as salt, sugar, herbs, spices, vitamins, and minerals, which are called added ingredients. These are added to foods either to improve the nutrient value or to add flavor.

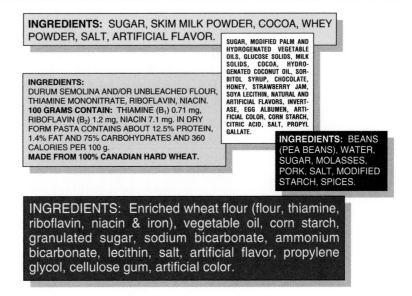

INGREDIENTS: SUGAR, SKIM MILK POWDER, COCOA, WHEY POWDER, SALT, ARTIFICIAL FLAVOR.

INGREDIENTS:
DURUM SEMOLINA AND/OR UNBLEACHED FLOUR, THIAMINE MONONITRATE, RIBOFLAVIN, NIACIN.
100 GRAMS CONTAIN: THIAMINE (B$_1$) 0.71 mg, RIBOFLAVIN (B$_2$) 1.2 mg, NIACIN 7.1 mg. IN DRY FORM PASTA CONTAINS ABOUT 12.5% PROTEIN, 1.4% FAT AND 75% CARBOHYDRATES AND 360 CALORIES PER 100 g.
MADE FROM 100% CANADIAN HARD WHEAT.

SUGAR, MODIFIED PALM AND HYDROGENATED VEGETABLE OILS, GLUCOSE SOLIDS, MILK SOLIDS, COCOA, HYDROGENATED COCONUT OIL, SORBITOL SYRUP, CHOCOLATE, HONEY, STRAWBERRY JAM, SOYA LECITHIN, NATURAL AND ARTIFICIAL FLAVORS, INVERTASE, EGG ALBUMEN, ARTIFICIAL COLOR, CORN STARCH, CITRIC ACID, SALT, PROPYL GALLATE.

INGREDIENTS: BEANS (PEA BEANS), WATER, SUGAR, MOLASSES, PORK, SALT, MODIFIED STARCH, SPICES.

INGREDIENTS: Enriched wheat flour (flour, thiamine, riboflavin, niacin & iron), vegetable oil, corn starch, granulated sugar, sodium bicarbonate, ammonium bicarbonate, lecithin, salt, artificial flavor, propylene glycol, cellulose gum, artificial color.

Figure 20.2   Many things are added to foods.

Federal government regulations control the food industry and the advertisement of foods. Ingredients must be listed on foods in order of decreasing mass. However, only categories of food additives are listed, and often standardized or individually packaged foods are not labeled at all.

Snack foods tend to contain more additives than most other foods. It is estimated that each of us consumes more than 2 kg of food additives each year. No one knows what the long-term effects of this type of consumption might be. However, it is known that certain groups of people, such as pregnant women, nursing mothers, small children, or people with kidney and liver diseases, might be at risk.

Table 20.1   Common food additives

Additive	Advantage	Disadvantage
food coloring	adds attractive colors	possible cause of hyperactivity in children
preservatives	prevent bacterial growth	possible allergen, may react to form suspected carcinogens (cancer-causing agents)
artificial sweeteners	low in calories	some are suspected carcinogens
flavor enhancers	create or increase taste	possible allergen, may be related to hyperactivity

Several other types of additives are used in foods. Some of these are bleaching agents to bleach doughs, thickeners to improve texture, emulsifying agents to prevent separation, and anticaking agents to improve flow.

Food additives are an important part of today's food industry. Unlike our ancestors, we do not all have time to grow our own food, harvest it, and spend time cooking and canning it. People often need meals in a hurry but want a variety of good-tasting, attractive foods. Additives help provide these qualities.

Additives have advantages and disadvantages. Information, testing, data, and time are needed to measure the value of these substances. Your knowledge of additives will help you to decide your food priorities.

Try These

1. Obtain five snack foods from a supermarket. Write down the list of ingredients. How many additives does each contain?
2. Compare the advantages and disadvantages of convenience foods and fresh foods.
3. What are food additives?
4. Give five examples of foods that contain added ingredients. Provide labels from the containers of these foods.

Extension

1. Chinese food often contains the additive monosodium glutamate (MSG). Some customers request no MSG in their food and some restaurants advertise that they do not use MSG. Why?

SCIENCE, TECHNOLOGY, & SOCIETY

Nitrates and Nitrites—Friend or Foe?

The nitrate anion (NO_3^-) and the nitrite anion (NO_2^-) are often found in water and food. Nitrates are a major component of fertilizer because "fixed" nitrogen in this form is essential for plant growth. Drinking water from farm wells will often contain small amounts of the nitrate anion because of fertilizer use. Soil bacteria can convert the nitrate anion to the nitrite anion. As a result, both can be found in drinking water.

Nitrates are used to preserve meats such as bacon and hot dogs.

The major source of nitrites ingested by humans is food **preservatives**. Sodium nitrite is used in processing ham, sausage, corned beef, bacon, bologna, and frankfurters. Nitrites are used to preserve the reddish color of the meat and to prevent the growth of botulism bacteria, which cause food poisoning.

One danger of the nitrite anion is that under acidic conditions or at high temperatures it reacts to produce nitrosamines. Nitrosamines are believed to be carcinogens. Nitrosamines can be formed when meats preserved with nitrites are cooked.

There is great debate about the nitrite anion. It prevents food poisoning but can form carcinogens. The evidence available so far suggests that the risk of botulism from eating preserved meats without nitrites would be far higher than the present risk of cancer. Nitrites continue to be used, but governments have limited the concentrations allowed.

Caution: If the product is to be eaten, all equipment must be sterilized.

Method

1. Mix a full test tube of glucose syrup and a full test tube of sucrose in a 250 mL beaker.
2. Heat gently until the mixture begins to boil. Stir constantly. Do not let the mixture burn.
3. Grease the aluminum foil with margarine. Add a small scoop of margarine to the beaker.
4. Heat and stir the solution until it thickens, starts to foam, and turns dark brown.
5. Add the protein, turn off the flame and stir.
6. As the mixture thickens, add the sodium bicarbonate and sodium chloride. Stir.
7. Pour the mixture onto the aluminum foil immediately. Let it cool before taste-testing.

Questions and Conclusions

1. What effect did the heat have on the sugar?
2. If your candy did not become brittle, what step was probably not carried out properly?
3. Name three sugars found in candy.
4. How does a cook control the crystals forming in candy?

20.8 ACTIVITY PREPARING ACETYLSALICYLIC ACID

The first pain-killing aspirin tablet was prepared and used by Felix Hoffmann, a junior chemist at the Bayer Company in Germany. Hoffmann's father had arthritis and was helped greatly by the drug.

Before Hoffmann discovered acetylsalicylic acid (ASA), salicylic acid had been used as an analgesic or painkiller, but it was very irritating to the stomach.

Hoffmann set out to modify salicylic acid in a way that would make it more soluble in stomach acids. This increased solubility let the analgesic pass through the stomach more rapidly and consequently it was not as irritating as previous analgesics.

Since 1899, ASA has been available to the public to reduce fever and pains from influenza, colds, muscular aches, arthritis, rheumatism, and tension. It may also benefit some people with heart problems, since it thins the blood.

There are two kinds of analgesic, peripheral and central. Peripheral analgesics, such as ASA and local anesthetics, act on the outer parts of the body. Central analgesics, such as morphine and general anesthetics, act directly on the brain.

Today, nearly everyone takes some form of ASA from time to time. However, recently Reye's syndrome has been associated with ASA. This disease has been found in rare instances among children and adolescents who have taken ASA when they had influenza or chicken pox.

In this activity you will prepare acetylsalicylic acid.

Materials

graduated cylinder
hot plate
pneumatic trough with ice
400 mL beaker
250 mL Erlenmeyer flask
glass stirring rod
medicine dropper
funnel
filter paper
distilled water
salicylic acid
acetic anhydride
sulfuric acid

Caution: Acids are corrosive and cause burns. Handle them with care. Wear goggles and an apron. If you spill any on yourself, rinse it off immediately with large amounts of cold water and have your partner tell your teacher.

Do not eat the ASA you prepare. School laboratory conditions do not meet pharmaceutical standards.

Method

1. Set up a hot water bath with 200 mL of water in a 400 mL beaker.
2. Put 1 g of salicylic acid into a 250 mL Erlenmeyer flask.
3. Add 3 mL of acetic anhydride to the salicylic acid.

4. Add five drops of sulfuric acid to the mixture from step 3 to acidify it.

5. Swirl the flask to mix the reactants and then heat the flask in the water bath at 80–90°C for at least 10 min.

6. Remove the flask from the water bath and allow it to cool to room temperature.

7. Place the flask in a trough of ice and add 15 mL of distilled water to the flask.

8. When the solution turns cloudy, scratch the inside of the flask with a stirring rod until crystals start to form.

9. When crystallization appears to be complete, filter the crystals. Wash them twice with small amounts of cold water and allow them to dry.

Questions and Conclusions

1. Why was salicylic acid not completely suitable as an analgesic?

2. Who discovered ASA?

Extension

1. Acetylsalicylic acid is the generic name or true name for the compound produced. Buying drugs that are generically named may be less expensive than buying brand name drugs. Comparison shop for the active ingredient, acetyl-salicylic acid. Report your findings to the class.

2. What is acetaminophen? What are its advantages and disadvantages in comparison with ASA?

20.9
DRUGS AND BODY CHEMISTRY

Drugs can be used to restore the body's chemical balance or to destroy it. When someone suffers a severe trauma in an accident, drugs can be used to restore that chemical balance. However, the same drugs upset the chemical balance in the body if they are abused. Several types of drugs affect our nervous system, the most common of which are discussed below.

Depressants

Depressants are usually barbiturates or tranquilizers, which depress the central nervous system and relieve tension. Barbiturates sedate or calm people in a manner that induces sleep, while tranquilizers sedate people without making them sleepy.

The general effect of all depressants is to slow the reflexes and make the user less alert. Large amounts may cause depression and emotional instability.

Alcohol is a commonly used depressant in many societies. Table 20.2 shows the effects of increasing levels of alcohol in the blood.

Table 20.2   Effects of different levels of blood alcohol

Number of Drinks (30 mL of 40% alcohol)	Blood Alcohol Percentage	Effects
<1	<0.03	sedating or tranquilizing
1–2	0.03–0.05	slowing of reaction time and reflexes
3–5	0.05–0.15	loss of coordination
6–8	0.15–0.20	reduced inhibitions, boisterous behavior
9–12	0.20–0.30	blurred vision, slurred speech, staggering
13–20	0.30–0.50	stupor, unconsciousness, coma, death

Hallucinogens

Hallucinogens create feelings of being in a dream world, but the effects of these drugs are unpredictable. The user's perception is altered and acts of violence, as well as hallucinations, can occur. The drug lysergic acid diethylamide, commonly known as LSD, is the most widely publicized hallucinogen. It has been known to cause deep depression, mental breakdown, and suicide.

Marijuana speeds up the heart rate and irritates the lungs. It can also affect the body's immune system. When the immune system is affected, the ability of white blood cells to respond to invading bacteria or viruses can be greatly reduced.

Marijuana is not addictive, but it can be habit-forming. Heavy users of this hallucinogen may become passive or even apathetic.

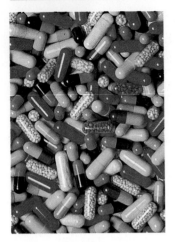

Stimulants affect the central nervous system.

Stimulants

Stimulants are most often amphetamines, commonly called "uppers" or "pep pills." Amphetamines affect the central nervous system. They produce a euphoric mood, and also increase heart rate, blood pressure, drive, and energy, thus postponing fatigue. They are not physically addictive but can be habit-forming.

Methamphetamine is a potent amphetamine known as "speed." Since the effects of this stimulant are prolonged, it is often necessary to counteract it with a barbiturate or heroin. For this reason, methamphetamine users often become heroin addicts.

Opiates

Opiates are the drugs most often linked to drug abuse. Morphine, codeine, and heroin are all opiates that have similar chemical structures. Morphine has the simplest structure and opium contains about ten percent morphine.

Opiates create euphoria and a feeling of tranquillity. Codeine is the least potent of the opiates and is used in some cough medicines. On the other hand, heroin is the most potent and most highly addictive narcotic known.

Caffeine and Nicotine

Caffeine stimulates the central nervous system and is an ingredient in coffee, tea, cocoa, and colas. It is also available in the form of tablets. It decreases drowsiness and increases alertness. Caffeine is not physically addictive but can be habit forming as shown by the number of people who are dependent on their cup of coffee first thing in the morning.

Nicotine in tobacco is addictive and use of smoking products containing this drug is linked to many cancer and heart disease deaths. Nicotine, however, is not the carcinogen or cancer-causing agent. The cause is the incomplete combustion products of the tobacco and the tobacco paper.

Questions

1. Give two reasons why a doctor might prescribe a drug.
2. List two reasons why drugs should not be abused.
3. What kinds of drugs are used as depressants?
4. What are the effects of depressants on the body?
5. At what level does alcohol start to slow reaction time?

6. At what level does alcohol produce intoxication?
7. What are the arguments for and against using alcohol?
8. Give two examples of hallucinogens.
9. Why are hallucinogens dangerous?
10. What kind of drug is most often used as a stimulant?
11. What is "speed"?
12. Why might an amphetamine be prescribed by a doctor?
13. Name three opiates.
14. Which opiate is the most dangerous? Why?
15. What dangers are associated with the drug nicotine?

20.10 ACTIVITY WHEN SHOULD DRUGS BE USED?

In this activity you will research the uses, abuses, advantages, and disadvantages of drugs and then debate the issue in class.

Method

1. Your teacher will divide the class into groups. One half of the students in a group should research and discuss reasons for the use of drugs. The other half should research and discuss the reasons why drugs should not be used.
2. The members of each group should be able to give reasons, data, and conclusive evidence in defense of their point of view.
3. Debate the topic within your group.
4. Prepare a list of reasons for drug use and a list of reasons against drug use.

OPPORTUNITY

Cosmetologist

Career opportunities in the beauty industry range from hairstyling and skin care to makeup artistry and product sales. The cosmetologist is a beauty service technician, involved in the cosmetic treatment of hair, skin, and nails. The cosmetologist advises customers on hair styles, cuts hair, and shampoos and colors hair. Before hair coloring can be done, the cosmetologist tests the sensitivity of the customer's scalp to determine whether there might be an allergic reaction. The beauty technician also gives scalp conditioning treatments for hygienic or remedial purposes.

Other services that cosmetologists perform include skin analysis and treatment for complexion problems, makeup application, the coloring of facial hair and lashes, straightening of hair, and cleaning and polishing of finger and toe nails. A knowledge of solutions and their properties is necessary.

Prospective candidates may specialize in a particular area of beauty services.

Exercises

Key Words

added ingredient
barbiturate
cosmetics
depressant
food additive
generic name
hallucinogen
opiate
preservative
stimulant
tranquilizer

Review

1. Define the term "cosmetics."
2. What types of cosmetics have you studied?
3. What is lanolin?
4. In what product is lanolin often used?
5. How does knowledge of a product's properties help in advertising it?
6. "All ingredients should be listed for all products." Comment on this statement.
7. Define the term "food additive."
8. Why might snack foods contain more than an average amount of additives?
9. What is meant by the term "confectionery industry"?
10. List the additives from the labels of ten food products.
11. Why is salt not considered an additive to butter?

Extension

1. What are the advantages of having the same cosmetic in different forms?
2. Research the controversy over generic and brand-name drugs.
3. Trace the process, costs, and testing procedures involved in developing a new drug.

CHAPTER 21
REDOX REACTIONS

Electrons are involved in all chemical reactions. This chapter studies reactions of ions with other substances, such as metals. In these reactions, electrons are transferred from one substance to another. While one substance loses electrons, another substance gains electrons.

One of the results of this type of transfer is the corrosion of metals. Corrosion and its prevention cost industries and consumers millions of dollars each year.

This chapter develops some of the many applications of the types of chemical reactions studied in Chapters 10 and 11.

Key Ideas

- Corrosion of metals takes place when metals lose electrons and become ions.

- Electron transfer occurs in a chemical reaction between an ion and a metal.

- When a substance loses electron(s) in a reaction it is oxidized.

- When a substance gains electron(s) in a reaction it is reduced.

- Oxidation and reduction occur simultaneously in a chemical reaction involving electron transfer.

- Electrons are transferred from the most active to the least active element.

- Some metals are more active than others. Hydrogen is used as a standard to compare the activities of metals.

- Corrosion and its prevention cost millions of dollars each year.

21.1 ACTIVITY
ACTIVE METALS AND CORROSION

Unless you work for an auto body repair shop, you would probably rather not think about corrosion of automobiles. Anyone who owns a car or truck fights corrosion because it is a major factor in the depreciation of a vehicle.

Corrosion is the loss of a metal's properties as a result of the action of air, water, and chemicals on it. In this process, a metal loses electrons and becomes a metal ion. This ion does not have the same properties as the metal and is often soluble, which causes it to dissolve from the surface of the metal. This action leaves more metal open to corrosion.

Active metals corrode easily. When electrons are rearranged in a chemical reaction, electrons are transferred from the most active to the least active element. An active metal corrodes because it gives up electrons to another ion.

Rain water may dissolve carbon dioxide or an industrial byproduct such as sulfur dioxide gas to form acidic solutions. Salt used on roads in winter also forms a solution. Each of these solutions contains ions that can react with unprotected iron such as that found in cars and trucks. Iron loses electrons and combines with oxygen in the air and becomes oxidized. The result of this chemical process is rust, a combination of iron oxides, **FeO** and **Fe$_2$O$_3$**.

In this activity you will compare the corrosion of iron and copper in the presence of hydrogen ions.

Corrosion causes depreciation of a car's value.

Materials

test tubes (7)	magnesium ribbon
test tube rack	copper(II) sulfate solution
iron strips or nails	iron(II) sulfate solution
copper strips or nails	dilute hydrochloric acid

Caution: Hydrochloric acid is corrosive and causes burns. Handle it with care. Wear goggles and an apron. If you spill any on yourself, rinse it off immediately with large amounts of cold water and have your partner tell your teacher.

Method

1. Number each of the test tubes from 1–7.
2. Add 10 mL copper(II) sulfate solution to test tubes 1 and 2.
3. Add 10 mL iron(II) sulfate solution to test tubes 3 and 4.
4. Add 10 mL of hydrochloric acid to test tubes 5 and 6.
5. Add an iron strip or nail to each of test tubes 1, 3, and 5.
6. Add a copper strip or nail to each of test tubes 2, 4, and 6.

The gold bracelet is not affected by hydrochloric acid, but the zinc-plated nail does react. What gas is formed?

7. Add 10 mL of hydrochloric acid to test tube 7.
8. Wrap an iron strip or nail with magnesium ribbon and put it into the test tube of acid.
9. Record your observations in chart form.

Questions and Conclusions

1. In which test tube(s) did you observe a reaction?
2. Which metal corrodes when placed in an acid?
3. Is the metal that corrodes in the acid more or less active than hydrogen?
4. Write the balanced chemical equation for the reaction.
5. What difference was there when you wrapped the nail with magnesium? What reason can you give for this difference?

21.2 ACTIVITY
COMPETITION
FOR ELECTRONS

The transfer of electrons can cause the corrosion of metals, such as those found in scrapyards.

When chemical bonds are formed, electrons are transferred or shared. Whenever substances come into contact with each other, they compete for each other's electrons. If electrons are transferred because of the competition, a chemical reaction occurs. Electrons are transferred from the most active to the least active element.

Metals and metallic ions always compete for electrons. Some metals lose electrons easily and others do not. By testing different combinations of metals and ions, you can rank the activity of the different metals from most active to least active. This pattern of activity is the basis for the activity series, redox reactions, and electrochemistry.

In this activity you will test different combinations of metals and ions. From the reactions observed, you will make up a list of metals to compare their reactivity or their ability to donate electrons. This is known as the activity series.

Materials

test tubes (4)	iron strip
test tube rack	copper strip
paper towels	zinc nitrate solution
emery paper	magnesium nitrate solution
magnesium strip	iron(II) nitrate solution
zinc strip	copper(II) nitrate solution

Method

1. Place 3 mL of copper(II) nitrate solution into each of four test tubes.
2. Clean the magnesium, zinc, iron, and copper strips with emery paper.
3. Add a small strip of each metal to a test tube of the copper(II) nitrate solution.
4. Observe each metal for several minutes.
5. Draw a chart similar to the one shown below and indicate where reactions occur. Draw a dash (—) if no reaction occurs.

	Cu	Zn	Mg	Fe
Cu^{2+}				
Zn^{2+}				
Mg^{2+}				
Fe^{2+}				

6. Remove the metals and dry them. Pour out the solutions as instructed by your teacher. Rinse the test tubes.
7. Repeat steps 1–6 with each of the other solutions, and record your observations.

Questions and Conclusions

1. Which metal had the fewest reactions?
2. Which metal had the most reactions?
3. What can you say about the ability of copper metal to donate electrons to
   a. zinc ions?   b. magnesium ions?   c. iron ions?
4. What can you say about the ability of magnesium metal to donate electrons to
   a. zinc ions?   b. copper ions?   c. iron ions?
5. Rank the metals from most active to least active.
6. How does the order in your answer to question 5 relate to the ability of each metal to donate electrons?
7. In which solution did the most reactions take place?
8. In which solution did the fewest reactions take place?
9. Rank the solutions in order from most active to least active.
10. How do the lists from questions 5 and 9 compare?
11. Make a general statement about the relative activity of a metal and its ion.

21.3
GAINING
AND LOSING
ELECTRONS

In Activity 21.2 you saw that zinc reacted with copper(II) nitrate. From this you learned that zinc is more active than copper. Zinc ions form and replace copper ions in solution.

When zinc metal and copper ions react, zinc metal loses two electrons to form a zinc ion, as follows:

$$Zn^0 \longrightarrow Zn^{2+} + 2\ e^- \tag{1}$$

The copper(II) ion gains two electrons from zinc to form solid copper metal.

$$Cu^{2+} + 2\ e^- \longrightarrow Cu^0 \tag{2}$$

If equations (1) and (2) are added together, the net result is the following chemical reaction:

$$Zn^0 \longrightarrow Zn^{2+} + 2\ e^- \tag{1}$$
$$Cu^{2+} + 2\ e^- \longrightarrow Cu^0 \tag{2}$$
$$\overline{Cu^{2+} + Zn^0 \longrightarrow Zn^{2+} + Cu^0} \tag{3}$$

Equations (1) and (2) are called half-reactions. Equation (3) is called the net equation. The nitrate ions do not participate in the reaction and because of this they are called spectator ions.

Magnesium metal reacts with a solution of silver nitrate. The following are the half-reactions and the net reaction.

$$Mg^0 \longrightarrow Mg^{2+} + 2\ e^- \tag{1}$$
$$2\ Ag^+ + 2\ e^- \longrightarrow 2\ Ag^0 \tag{2}$$
$$\overline{2\ Ag^+ + Mg^0 \longrightarrow 2\ Ag^0 + Mg^{2+}} \tag{3}$$

However, a similar reaction does not occur when silver is placed in a solution of magnesium nitrate. Silver metal is less active than magnesium metal and therefore cannot replace the magnesium ions present in the solution.

Try These

1. Write the half-reactions for the reaction between
   a. iron metal and copper ions
   b. magnesium metal and iron ions
2. a. Write the net equation for the following half-reactions:

$$Zn^0 \longrightarrow Zn^{2+} + 2\ e^-$$
$$Fe^{2+} + 2\ e^- \longrightarrow Fe^0$$

   b. If the spectator ion is SO_4^{2-}, what would be the overall single displacement reaction?
3. Would you expect copper metal to react if it was placed in a zinc nitrate solution? Explain your answer.

21.4
REACTIONS WITH ELECTRON TRANSFER— REDOX REACTIONS

In Chapters 10 and 11 you studied four general types of reactions. These were direct combination or synthesis, decomposition, single displacement, and double displacement.

The first three types, direct combination, decomposition, and single displacement reactions, often involve electron transfer. For example, the following is a direct combination reaction that involves electron transfer:

$$2 \, Mg + O_2 \longrightarrow 2 \, MgO$$

Each magnesium atom in the reaction loses two electrons to become a magnesium ion:

$$Mg^0 \longrightarrow Mg^{2+} + 2 \, e^- \qquad (1)$$

Each oxygen atom in the reaction gains two electrons to become an oxygen ion:

$$O^0 + 2 \, e^- \longrightarrow O^{2-} \qquad (2)$$

The net equation is:

$$2 \, Mg^0 \longrightarrow 2 \, Mg^{2+} + 4 \, e^- \qquad (1)$$
$$O_2 + 4 \, e^- \longrightarrow 2 \, O^{2-} \qquad (2)$$
$$\overline{2 \, Mg + O_2 \longrightarrow 2 \, MgO \qquad (3)}$$

Whenever a metal combines with oxygen to form an oxide, the metal is said to be oxidized. In this type of reaction the metal loses electrons. The name for this type of half-reaction is oxidation. The name oxidation is used even when oxygen is not involved. Oxidation involves the loss of one or more electrons and results in an increase in charge number.

When one substance loses electrons, another substance gains them. The name of the half-reaction in which the substance gains electrons is reduction. Reduction involves the gain of one or more electrons and results in a decrease in charge number.

Oxidation and reduction half-reactions occur together. The total net equation is called an oxidation-reduction reaction or more commonly, a redox reaction.

In the reaction between zinc and sulfur, zinc loses electrons and is oxidized, while sulfur gains electrons and is reduced.

$$Zn^0 \longrightarrow Zn^{2+} + 2 \, e^- \qquad \text{(oxidation)} \quad (1)$$
$$S^0 + 2 \, e^- \longrightarrow S^{2-} \qquad \text{(reduction)} \quad (2)$$
$$\overline{Zn^0 + S^0 \longrightarrow ZnS \qquad \text{(redox)} \quad (3)}$$

The term "oil rig" is a memory aid that can be used to recall the definitions of oxidation and reduction. "Oil" stands for "**o**xidation **i**s **l**oss" and "rig" stands for "**r**eduction **i**s **g**ain."

Oxidation
Is
Loss

Reduction
Is
Gain

Figure 21.1   The words OIL RIG can be used as a memory aid.

Try These

1. Which substances are oxidized and which are reduced in the following reactions?
   a. $4\ Na + O_2 \longrightarrow 2\ Na_2O$
   b. $2\ CuO \longrightarrow 2\ Cu + O_2$
   c. $Zn + FeCl_2 \longrightarrow ZnCl_2 + Fe$
2. a. Which type of half-reaction is represented by the following?
   $Zn^0 \longrightarrow Zn^{2+} + 2\ e^-$
   b. Does the reactant, zinc metal, gain or lose electrons in this half-reaction?
3. Which of the four general types of chemical reaction studied do not involve redox reactions?

21.5 ACTIVITY
THE ACTIVITY SERIES AND HYDROGEN

In Activity 21.2 you saw that active metals lose electrons and, as a result, are oxidized. Active metals react with acids, such as hydrochloric acid, causing the release of hydrogen gas from the acid.

In this activity you will compare the relative activity of some metals to that of hydrogen (a standard) and rank the metals in an activity series.

Because some metals are so active and often react violently with acid, they are better tested with water. These reactions (Parts A and B) will be demonstrated by your teacher. You will do Part C.

Materials

test tubes (5)
test tube rack
150 mL beakers (2)
emery paper
aluminum
zinc
iron
copper
silver
hydrochloric acid
copper(II) sulfate solution
silver nitrate solution

Caution: Silver nitrate and hydrochloric acid are corrosive and cause burns. Handle them with care. Wear goggles and an apron. If you spill any on yourself, rinse it off immediately with large amounts of cold water and have your partner tell your teacher.

Method

Parts A and B are Teacher Demonstrations

Part A

1. Your teacher will react a small piece of each of lithium, sodium, potassium, and calcium with water. Note how each metal is stored and the safety precautions that are used. Record your observations. Compare the relative activity of each metal.

Part B

1. Your teacher will react magnesium with steam.

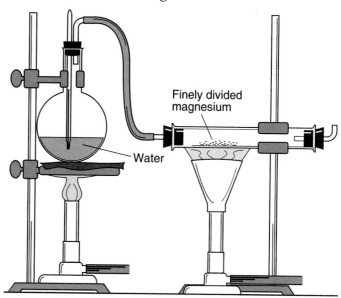

Figure 21.2 The reaction of steam with finely divided magnesium

Part C

1. Use emery paper to clean the surfaces of samples of aluminum, zinc, iron, copper, and silver.
2. Place 5 mL of hydrochloric acid into each test tube.
3. Place samples of each of the metals into each test tube. Note the reactivity of the metals by observing how quickly hydrogen gas is released. Record your observations in a chart.
4. Place a strip of silver into a beaker of copper(II) sulfate solution and a strip of copper into a beaker of silver nitrate solution. Let them stand for a few minutes. Record your observations.

Questions and Conclusions

1. Were there metals that did not react with the acid to produce hydrogen? Which one(s)?
2. Which was the least active metal used?
3. Which metals were more active than hydrogen?
4. Use the results to rank the elements tested, from most active to least active. (Note: Calcium and sodium are close in activity, but calcium is more active.)
5. When hydrogen gas is produced from an acid, what half-reaction is taking place?

Where do you think metals that are used in coins would be found on the activity series?

21.6 ACTIVITY HYDROGEN AND ITS PROPERTIES

From the activity series you developed in Activity 21.5, you know that many metals are more active than hydrogen. The following is a list of the relative activity of some metals.

The Activity Series of Some Elements

1. lithium
2. potassium
3. calcium
4. sodium
5. magnesium
6. aluminum
7. zinc
8. chromium
9. iron
10. nickel
11. tin
12. lead
13. **HYDROGEN**
14. copper
15. mercury
16. silver
17. platinum
18. gold

You can use the information in the activity series to predict which metals will release hydrogen gas from an acid by a single displacement reaction. Recall that those metals that reacted with water in Activity 21.5 also reacted vigorously with acid.

Hydrogen was discovered in 1766 when Henry Cavendish collected some gas from the reaction of iron and sulfuric acid. He called the gas "inflammable air" because it burned. Since the only product of this combustion reaction was water, another scientist, Antoine Lavoisier, suggested that the gas be named "hydrogen" from the Greek word for "water producer."

Any metal more active than hydrogen can be used to produce hydrogen. In this activity you will produce and collect hydrogen by reacting zinc with hydrochloric acid. You will then study some of the properties of hydrogen gas.

Materials

retort stand

utility clamp

pneumatic trough

rubber tubing

250 mL beaker

test tube rack

large test tube

small test tubes (4)

small stoppers (4)

rubber stopper with glass tubing

wooden splint

mossy zinc

dilute hydrochloric acid

Caution: Hydrogen mixed with air is explosive. Do not have open flames near the apparatus while it is in operation.

   Hydrochloric acid is corrosive and causes burns. Handle it with care. Wear goggles and an apron. If you spill any on yourself, rinse it off immediately with large amounts of cold water and have your partner tell your teacher.

Method

1. Set up the apparatus as shown below.

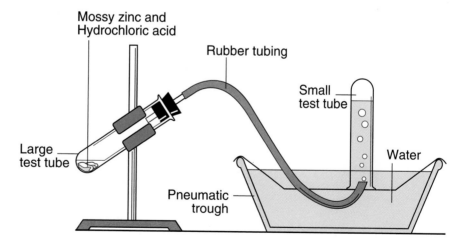

Figure 21.3   Apparatus for producing hydrogen gas

2. Fill four small test tubes with water and invert them in the trough. Do not let any water escape.
3. Put several lumps of zinc into the large test tube. Add 15 mL of hydrochloric acid and insert the stopper with glass tubing.
4. Let the reaction continue for 30 s before putting the end of the tubing under the test tube in the trough.
5. Fill four test tubes with hydrogen. Remove each test tube, stopper it, and place it in a test tube rack.
6. When four test tubes have been collected, remove the stopper from the large test tube. Carefully pour out the acid and rinse the zinc with water.
7. Observe one of the test tubes. Note the color of hydrogen gas. Remove the stopper from the test tube and note the odor of the hydrogen gas.
8. Let the test tube stand open for 2 min and test with a burning splint by inserting it into the mouth of the test tube. Record your observations.
9. Insert a burning splint into the mouth of a second test tube. Record your observations.
10. Place a test tube of hydrogen gas mouth downward into a beaker of water. Swirl it around gently. Record your observations.
11. Hold one test tube upside down and put a burning splint well into the test tube. Remove it slowly. Note any difference between burning at the mouth of the tube and inside the tube. Observe the walls of the tube.

Questions and Conclusions

1. How would you describe the following properties of hydrogen gas?
   a. color
   b. odor
   c. solubility in water
   d. density compared with air
2. What is the product of the combustion of hydrogen?
3. Write the balanced chemical equation for the burning of hydrogen.
4. Write the balanced chemical equation for the reaction between zinc and hydrochloric acid.
5. What property of hydrogen makes it dangerous?
6. Does hydrogen support combustion? Explain your answer.
7. What two properties of hydrogen are illustrated in the photo on page 387?

SCIENCE & SOCIETY

Hydrogen

Hydrogen is the simplest and most abundant element known. It is a major constituent of all stars. On earth, most hydrogen is found in the form of compounds, namely water and acids, and in living organisms as carbohydrates.

Hydrogen has a low density, 0.09 kg/m^3, so that hydrogen has buoyancy or "lift ability." This physical property was used in dirigibles. Hydrogen has important chemical properties as well. Its reactivity resulted in the *Hindenburg* disaster. The chemical reactivity of hydrogen, especially in its reaction with oxygen, has many useful applications. It can be used in fuel cells and it is also capable of generating enough thrust to be used as a rocket fuel.

The *Hindenburg*, a hydrogen dirigible, crashed in flames in New Jersey in 1937, killing 36 people.

Because of its explosive reaction with oxygen, many safety precautions are necessary. The space shuttle *Challenger* exploded on January 28, 1986 when the liquid hydrogen fuel tank ruptured after a seal was damaged.

Aside from its useful physical and chemical properties, hydrogen has nuclear potential as well. One isotope of hydrogen, tritium (^3_1H), can fuse with another isotope of hydrogen, deuterium (^2_1H), to form helium and a neutron. This reaction releases about seven million times as much energy as the equivalent hydrogen-oxygen chemical reaction. This nuclear reaction is used in hydrogen fusion bombs. Soon we may be able to control it and use it in nuclear fusion reactors to produce electricity. The consequences for society are extreme in both benefits and risks.

Producing hydrogen is still relatively expensive. Electrolytic cells produce hydrogen and oxygen by passing electricity through water. As these cells become more efficient and less expensive, the use of hydrogen as an everyday fuel will become more than a dream.

Hydrogen bomb explosions release large amounts of energy and radioactivity.

21.7 CORROSION PREVENTION

Iron is the metal used in most of today's construction. Because it corrodes or rusts easily, iron needs protection from substances in the atmosphere. Corrosion-prevention treatments for metals are very costly. However, the damage caused by corrosion of unprotected metal costs millions of dollars each year. Think of the cost of replacing cars alone.

There are several ways to protect iron from corrosion. The following are some anticorrosive measures.

Painting: This is the most common way of protecting iron. The paint coats it and keeps water and air from contacting the metal. However, chips and scratches in the paint can expose the iron and leave it open to rusting.

Galvanizing or Tin Plating: Iron is galvanized by dipping it into melted zinc. When iron is tin plated, melted tin is used. A thin layer of tin or zinc protects the iron from corrosion.

Alloying: Under extreme conditions coatings do not always give enough protection. Alloys, which are blends of metals, give long-lasting protection. Stainless steel is a familiar example of an alloy. It is a blend of iron, chromium, and nickel.

Electroplating: Electroplating iron involves applying a coating of a metal such as silver, chromium, nickel, or tin. Silver-plated dinnerware is an example of electroplated metal.

Protection Using an Active Metal: To prevent the corrosion of ships' hulls, magnesium bars are attached to the hulls of ships. Magnesium is more active than iron, so the magnesium bars will corrode in seawater before the iron hull will.

SCIENCE & TECHNOLOGY

Alloys

An alloy is a substance that has metallic properties and is composed of two or more elements, at least one of which is a metal. Alloys are designed for particular applications and the property of the alloy is suited to its use.

For airplanes, aluminum metal is useful because of its low density, but pure aluminum is not strong enough. To strengthen it, an alloy is produced by adding small amounts of elements such as copper, magnesium, zinc, or silicon.

Aluminum alloys are used to build airplanes.

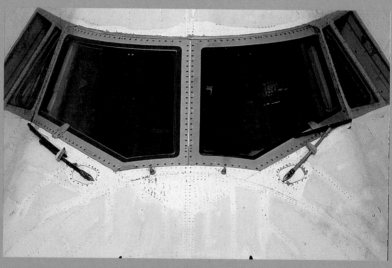

Pure gold is too soft for jewelry. Gold jewelry is always made from an alloy of gold. Copper is the metal most commonly added but silver, nickel, and zinc are also used. The gold color and luster are not affected.

Pure gold is 24 carat gold. Eighteen carat gold means that the percentage of gold present is 18/24, or 75% gold. The percentage gold is the "carat value" divided by 24 (for pure gold) and multiplied by one hundred.

Sterling silver is also an alloy. It contains 92.5% silver and 7.5% copper.

OPPORTUNITY

Craft Jeweler

Being a craft jeweler provides independence and an opportunity to be creative. In order to become a craft jeweler you need to know about metals, their malleability and thermal conductivity, and their alloys. Skills in cutting, bending, shaping, and soldering are all fundamental to the creative process.

To become a craft jeweler, you can study courses at college. There are opportunities for jobs at jewelry factories and large jewelry stores as well as the opportunity to set up your own shop.

Exercises

Key Words

activity series

corrosion

electron transfer

half-reaction

net equation

oxidation-reduction
   reaction

redox reaction

reduction

relative activity

rust

spectator ion

standard

Review

1. What is meant by "corrosion"?
2. a. Does the more active or less active metal lose electrons more easily in the competition for electrons?
   b. How is the relative activity of a metal related to the activity series?
3. What is the function of spectator ions in a chemical reaction?
4. Define "oxidation" and "reduction."
5. In the following reaction, which substance is oxidized?
$$Mg^0 + S^0 \longrightarrow Mg^{2+} + S^{2-}$$
6. How did hydrogen get its name?
7. Describe two physical and two chemical properties of hydrogen.
8. Using the activity series from Section 21.6, which of the following reactions would occur?
   a. **Zn** is oxidized/**Ni** is reduced
   b. **Cu** is oxidized/**Na** is reduced
   c. **K** loses electrons/**Na** gains electrons
   d. **Hg** is oxidized/**Ca** is reduced
9. What is rust?
10. List five ways to prevent corrosion.
11. a. Give the net equation for the following half-reactions:
$$Zn^0 \longrightarrow Zn^{2+} + 2\ e^-$$
$$Cu^{2+} + 2\ e^- \longrightarrow Cu^0$$
   b. If the spectator ion is sulfate, SO_4^{2-}, write the balanced chemical equation for the above reaction.

Extension

1. Use the library resource center to research one of the methods used to prevent corrosion and report your findings to the class. If possible, test the method you have chosen.
2. Find out how metals are recycled in your municipality or region. Report on how the process works and what it costs. Does recycling metals make or lose money for the taxpayer?

CHAPTER 22
ELECTROCHEMISTRY

In Chapter 21 you learned that electrons may be transferred during chemical reactions. By directing the way a reaction will take place, we can make electrons flow from one substance to another through a wire. This flow of electrons is called electricity and is the chemical principle of electrochemical cells.

We can also reverse the procedure and use electricity to cause a chemical reaction to occur. Auto workers do this when they electroplate bumpers and other car parts with chromium.

This type of chemistry is called electrochemistry and has many important applications in industry.

Key Ideas

- Batteries use chemical reactions to produce electricity.
- Electricity can be used to cause a chemical reaction to occur.
- Electrolysis is the name for the process that uses electricity to cause redox reactions to occur that do not normally occur.
- Electrolysis is the basis of the electroplating industry.

$$Zn \longrightarrow Zn^{2+} + 2\ e^- \qquad E^0 = 0.76\ V$$

$$Cu^{2+} + 2\ e^- \longrightarrow Cu^0 \qquad E^0 = 0.34\ V$$

$$Cu^{2+} + Zn^0 \longrightarrow Cu^0 + Zn^{2+} \qquad E^0_{cell} = 1.10\ V$$

Try These

1. What is a standard?
2. At which electrode in an electrochemical cell does oxidation occur?
3. Calculate the net cell voltage for each of the following cells:
   a. magnesium-silver   b. iron-copper   c. zinc-silver
4. Which substance is oxidized and which is reduced in each of the cells named in question 3?

22.4

BATTERIES

One of the most common batteries is the lead acid storage battery, which is actually six electrochemical cells connected together in a series. This type of battery is used in a car to supply the electricity needed to start the car engine. The anodes are lead, the cathodes are lead dioxide, and the electrolyte that supplies ions in solution is sulfuric acid.

The dry cell is another familiar cell. It is commonly called a battery and is made up of a single cell. The container is made of zinc, which acts as the anode. The center post, on the inside of the cell, is made of carbon and acts as the cathode. In the alkaline dry cell, the voltage is similar to the common dry cell, but it can work for a longer time, give more electricity, and still maintain its voltage.

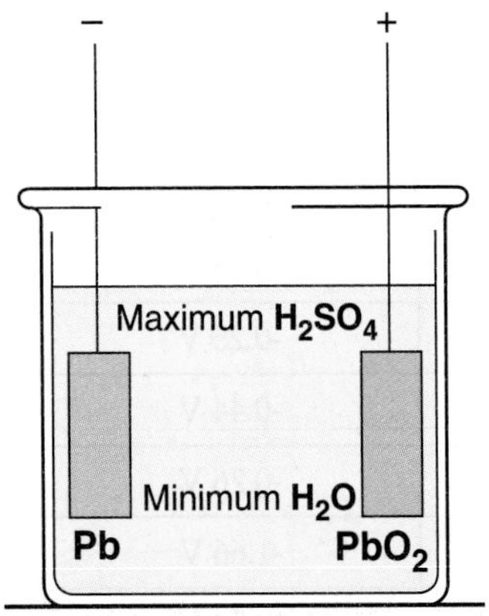

Figure 22.3
The state of the electrodes in one cell of a charged lead acid storage battery

Try These

1. The overall reaction for a car battery is

   $$Pb + PbO_2 + 2\ H_2SO_4 \longrightarrow 2\ PbSO_4 + H_2O$$

   What happens when a car battery is recharged?
2. The voltage of one cell in a car battery is approximately 2 V. What is the total voltage of a six-cell car battery?
3. Identify the cathode and anode of a common dry cell.

22.5 ACTIVITY
COMPARING ELECTROCHEMICAL CELLS

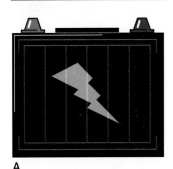

A

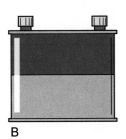

B

C

D

E

Figure 22.4
Some common electrochemical cells.
(See Table 22.3)

Cells and batteries, which are combinations of cells, come in all shapes and sizes and power everything from cars to watches. In this activity you will compare the properties of several common cells.

Method

1. Examine Figure 22.3 and identify the anode, the cathode, and the electrolyte.
2. Compare the cells in Figure 22.4 to Table 22.3. Match the letter of a cell to the following uses. Explain your choices.
   a. toy car   b. hearing aid   c. lantern flashlight

Table 22.3   Features of some common cells

Cell	Type	Size	Voltage	Energy
A	lead acid storage	large	12 V	high
B	dry cell Zn-C	large	6 V	high
C	nickel-cadmium	large	7.2 V	high
D	dry cell Zn-C	small (AA)	1.5 V	low
E	nickel-cadmium	small (button)	1.2 V	low

Questions and Conclusions

1. Define the terms "oxidation" and "reduction."
2. At which electrode does each half-reaction in Figure 22.3 occur?
3. What causes the differences in voltage in the cells shown in Figure 22.4?
4. Which of the following factors do you think is most important for a lantern flashlight cell? Explain your choice.
   a. size   b. cost   c. voltage   d. length of usefulness
5. Which of the following factors do you think is most important for a hearing aid cell? Explain your choice.
   a. size   b. cost   c. voltage   d. length of usefulness
6. Which of the following factors do you think is most important for a toy car cell? Explain your choice.
   a. size   b. cost   c. voltage   d. length of usefulness

22.6
THE FUEL CELL

The batteries or cells you have looked at so far are limited because they wear out. A perfect cell would be one in which the chemicals are always fresh. This type of cell does exist and is called a fuel cell. In a fuel cell, the chemicals are renewed and the products of the reaction are removed. The hydrogen-oxygen fuel cell is an example of such a cell. Hydrogen is brought to the anode and oxidized, and oxygen is brought to the cathode and reduced.

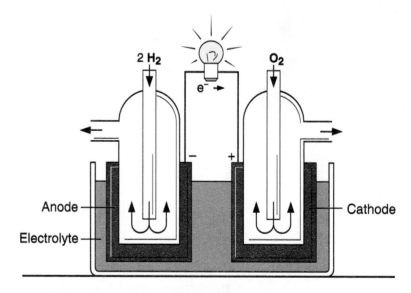

Figure 22.5   A hydrogen-oxygen fuel cell and its main parts

In these cells, hydrogen and oxygen provide electricity and produce water as a byproduct. The future use of fuel cells looks promising in cases where cost is not very important.

Some familiar examples of the uses for fuel cells to date are in the Apollo space program, submarines, electric cars, communication systems, and hospitals.

The advantages are that fuel cells are efficient, cause little pollution, and are a reliable source of power. The disadvantages are that they are expensive and the electrodes wear out quickly.

SCIENCE, TECHNOLOGY, & SOCIETY

Pacemakers

Many people have abnormal heart rhythms (arrhythmia). Their heartbeat may be too fast or too slow or may fluctuate suddenly. For people with a condition called heart block, the normal impulses controlling the heartbeat are interrupted.

These conditions can be treated by implanting a small electronic device into the body of the patient. This device, called a pacemaker, is connected to the heart and transmits an electric impulse to the heart, stimulating it to beat in a regular rhythm. The first implantable pacemaker was developed by Dr. Wilfred Gordon Bigelow at the University of Toronto.

Most pacemakers use a lithium battery that lasts from five to fifteen years. Some pacemakers have a rechargeable battery; others use a nuclear-powered battery.

22.7 ELECTROLYSIS— REVERSING THE PROCESS

Through the competition for electrons, chemical reactions can be used to produce electricity. When the process is reversed, electricity is used to cause chemical reactions. This process is called electrolysis. Electrolysis causes redox reactions to occur that would not normally occur.

The source of electricity for electrolysis is a power supply, such as an electrochemical cell or battery. The cell where electrolysis occurs is called an electrolytic cell. A familiar example of this is the recharging of a car battery. After extended use, as it is discharged, a car's battery loses voltage because the sulfuric acid has been used up and, in its place, water has been produced. However, the sulfuric acid can be restored by using a power source to reverse the reaction and recharge the battery.

$$PbO_2 + Pb + 2\,H_2SO_4 \underset{\text{recharge}}{\overset{\text{discharge}}{\rightleftharpoons}} 2\,PbSO_4 + 2\,H_2O$$

Electroplated silverware

Electrolysis has many useful applications, for example, electroplating, which includes silverplating cutlery and chromeplating auto parts.

The electrolyte in an electrolytic cell can be a molten ionic substance or an ionic solution. In either case, the electrodes do not take part in the reaction except to transfer electrons.

When the electrolyte is a molten (melted) salt, such as sodium chloride, the reaction is as follows:

$$2\,Na^+ + 2\,e^- \longrightarrow 2\,Na^0 \qquad \text{(Reduction)}$$

$$2\,Cl^- \longrightarrow Cl_2 + 2\,e^- \qquad \text{(Oxidation)}$$

$$\overline{2\,Na^+ + 2\,Cl^- \longrightarrow 2\,Na^0 + Cl_2}$$

When the electrolyte is a solution of sodium chloride, the reaction is slightly different because water is present and takes part in the reaction. Water is reduced at the cathode instead of sodium because less energy is needed to reduce water. In this type of cell, ions carry the electrons from one electrode to the other to complete the circuit.

$$H_2O + 2\,e^- \longrightarrow H_2 + 2\,OH^- \qquad \text{(Reduction)}$$

$$2\,Cl^- \longrightarrow Cl_2 + 2\,e^- \qquad \text{(Oxidation)}$$

$$\overline{H_2O + 2\,Cl^- \longrightarrow H_2 + Cl_2 + 2\,OH^-}$$

In the following activity you will take a closer look at electrolysis and electrolytic cells.

Try These

1. Define "electrolysis."
2. What is the role of the electrodes in electrolysis?
3. What can be used as a power source for electrolysis?
4. What is the role of the electrolyte in electrolysis?

22.8 ACTIVITY
ELECTROLYTIC
CELLS

In Part A of this activity you will use a power supply to pass electricity through water. Dilute sulfuric acid is added to the water to make it a better electrolyte. The reaction that occurs produces two gases. You will perform tests to identify the gases and determine at which electrode each gas was formed.

In Part B you will use electrolysis to electroplate an object.

Materials

Hoffmann apparatus
   (**or** J-shaped electrolysis electrodes (2)
      test tubes (2)
      400 mL beaker)
6–12 V power supply
dry cell
conducting wires (2)
glass rod
400 mL beaker
emery paper
wooden splint
medicine dropper
object to be plated (e.g., a metal key or an iron strip)
copper strip
dilute sulfuric acid
copperplating solution
phenolphthalein indicator

Caution: Dilute sulfuric acid and copperplating solution are corrosive and cause burns. Handle them with care. Wear goggles and an apron. If you spill any on yourself, rinse it off immediately with large amounts of cold water and have your partner tell your teacher.

Method

Part A

1. Set up the apparatus as shown in Figure 22.6 or Figure 22.7 or as instructed by your teacher.
2. Fill the Hoffmann apparatus or beaker and inverted test tubes with water that has had sulfuric acid added to it. Add a few drops of phenolphthalein indicator.
3. Connect the power supply. Note the positive and negative terminals. Let the apparatus run and record your observations. Meanwhile, prepare your apparatus for Part B.

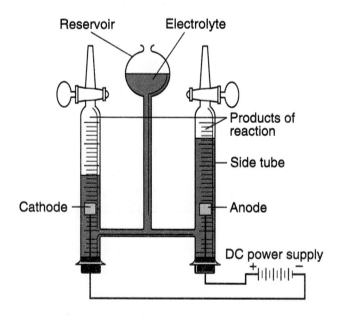

Figure 22.6   Hoffmann apparatus

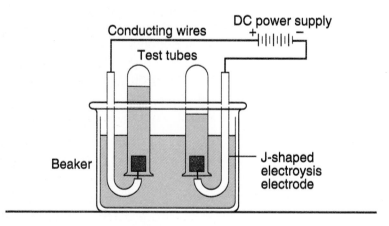

Figure 22.7   Alternate set-up

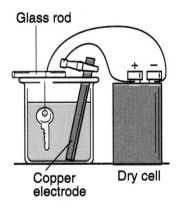

Glass rod

Copper electrode

Dry cell

Figure 22.8
Electroplating a key

4. When a test tube full of gas (or its equivalent) has been collected, disconnect the power supply.
5. If using the Hoffmann apparatus, fill an inverted test tube with gas by opening the valve above the anode.
6. Repeat step 5 at the cathode.
7. Remove the test tube of gas collected above the cathode or negative terminal. Keep the test tube inverted. Test the gas with a burning splint and record your observations.
8. Remove the test tube of gas collected above the anode or positive terminal. Keep the test tube inverted. Test the gas with a glowing splint and record your observations.

Part B

1. Fill the beaker with the copperplating solution.
2. Clean the object to be plated with the emery paper. Suspend the object in the copperplating solution using a conducting wire connected to the negative terminal, as shown in Figure 22.8.
3. Connect the copper strip to the positive terminal. Record your observations. After the object is well plated, disconnect the power supply.
4. Remove the plated object and rinse it.
5. The plated object can be polished with a fine abrasive.

Questions and Conclusions

1. What gas is produced at the anode in Part A?
2. What gas is produced at the cathode in Part A?
3. The overall equation for the electrolysis of water is

$$2\,H_2O \longrightarrow 2\,H_2 + O_2$$

The half-reactions are

$$4\,H^+ + 4\,e^- \longrightarrow 2\,H_2$$

$$2\,H_2O \longrightarrow O_2 + 4\,H^+ + 4\,e^-$$

At which electrode does each of these half-reactions occur?
4. What did the pink color of the phenolphthalein show?
5. Why was some sulfuric acid added to the water?
6. Write the half-reaction that takes place at each electrode during the copperplating.
7. An object that is to be silverplated must be connected at which electrode of an electrolytic cell? Why?
8. Give one other example of electroplating.

22.9
ELECTROPLATING

Electroplating can be used as a method to prevent corrosion and at the same time to beautify an object. A set of sterling (92.5%) silver tableware is extremely expensive, so many people today buy silverware that has been silverplated.

Knives, forks, and spoons are plated electrically with silver. They appear identical to solid silver but are much less expensive. For the same reason, gold chains may also be plated with gold rather than being made of solid gold.

Chrome-plated car parts have long been a familiar sight. However, as the technology of plastics has improved, many chrome parts have been replaced. For the car enthusiast, restoring a classic automobile often means replating bumpers, moldings, and other corroded parts. Triple plating can be used. This means the object is plated three times using copper first, then nickel, and chromium last.

From Activity 22.8 you learned that reduction occurs at the cathode of the electrolytic cell. The object to be plated with metal must be connected so that it acts as the cathode. Many factors determine the quality of the finish on the object. The cleanliness of the object, its finish, and the current and voltage used in the power source all affect the product.

A vintage automobile has many chrome-plated parts.

Try These

1. For an object to be electroplated, which electrode must it be?
2. What is triple plating?
3. Why is chromium or chrome plating so useful for cars?
4. What are the advantages of silver or gold plating?
5. Describe the process for plating a lead object with copper.

22.10 ACTIVITY INDUSTRIAL APPLICATIONS OF ELECTROLYSIS

Copper is refined by electrolysis. Blister copper, which is impure copper, is the anode, pure copper is the cathode, and copper(II) sulfate is the electrolyte. During electrolysis, copper ions are reduced at the cathode, increasing the mass of the cathode. Eventually the anode is used up and must be replaced. A slush of the impurities of blister copper, including platinum, gold, and silver, forms below the anode and can be collected. Pure copper is plated at the cathode. By separating the impurities from the blister copper, the metal, copper, can be purified.

Copperplating

A cell with molten sodium chloride as the electrolyte is used to make pure sodium metal. The industrial version of this cell is called a Down's cell.

Pure aluminum is also produced by electrolysis; this is called the Hall-Héroult process. Bauxite (Al_2O_3) is decomposed and molten aluminum is produced at the cathode and collected from the container. Aluminum is a useful metal and is used in cans, airplanes, and electric power lines.

In this activity you will investigate and report on one industrial use of electrolysis.

Method

1. Research to find some of the industries that use electrolysis.
2. Choose an industry that uses electrolysis and that interests you.
3. Use reference materials to find out how electrolysis is applied in that industry.
4. Write a complete report (1000 words) on electrolysis and the use of redox reactions in the industry that you have chosen.

Questions and Conclusions

1. When copper, sodium, and aluminum are being produced by electrolysis, at which electrode do they collect?
2. What type of half-reaction occurs to produce each of the metals in question 1?

Exercises

Key Words

battery
cathode
dry cell
electrochemical cell
electrochemistry
electrode
electrolytic cell
electroplating
fuel cell
half-cells
net cell voltage
salt bridge
voltage

Review

1. What is electricity?
2. Draw an electrochemical cell showing its main parts.
3. What reaction occurs at the anode of an electrochemical cell?
4. What amount of voltage is produced in an **Al-Fe** cell?
5. Which metal will be the anode in the **Al-Fe** cell?
6. What is a battery?
7. How is a fuel cell different from a battery?
8. List three advantages and three disadvantages of fuel cells.
9. Name three places where fuel cells are used.
10. Define "electrolysis."
11. What is the difference between an electrochemical cell and an electrolytic cell?
12. Give an example of electrolysis at work.
13. Why is chrome plating used on some car parts?
14. Describe the Hall-Héroult process for making aluminum.

Extension

1. Report on the uses of electrolysis in metallurgy.
2. Obtain information from a manufacturer about factors to be considered in selecting dry cells.
3. Prepare a report on the future use of fuel cells.
4. In what ways are a dry cell and an alkaline dry cell different?
5. Report on a career involving electrochemistry.

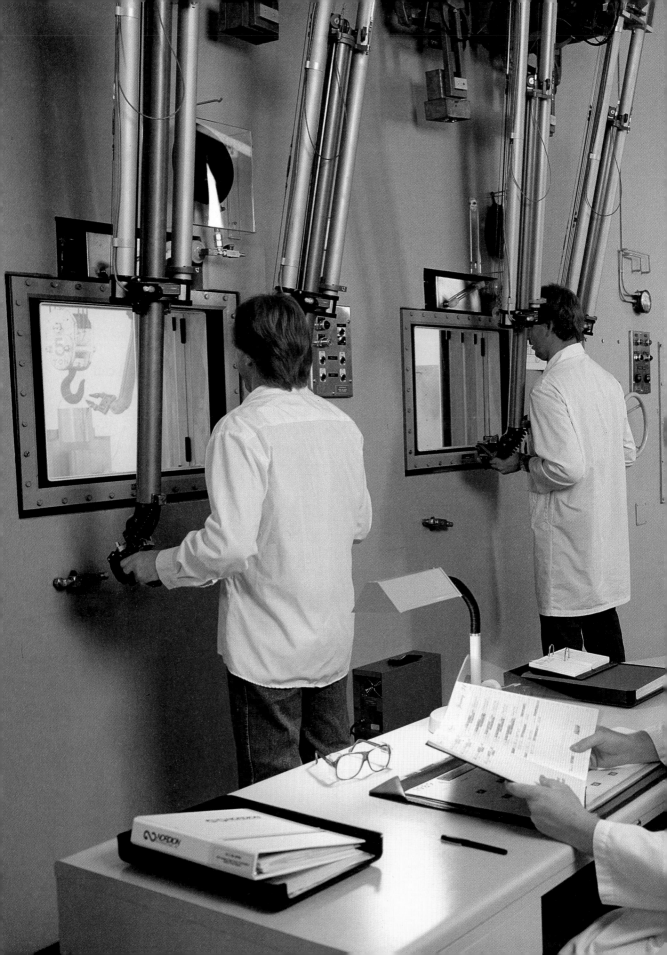

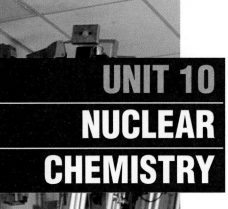

UNIT 10
NUCLEAR
CHEMISTRY

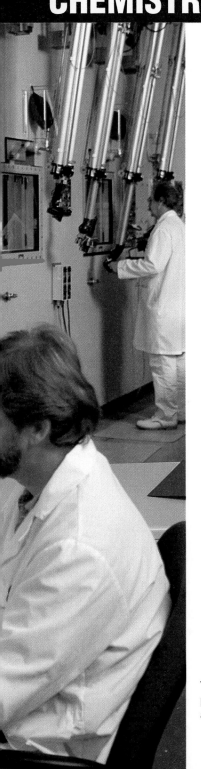

CHAPTER 23
RADIOACTIVITY

The discovery and understanding of the nuclear structure of the atom in the 20th century have resulted in some technological advances that have been very beneficial to society. These have made possible the prospect of vast supplies of nuclear energy and the use of radioisotopes in research, food preservation, and medicine.

However, many controversial issues have arisen. For example, use of nuclear weapons led to the destruction of Hiroshima and Nagasaki, causing the loss of many human lives and much misery. The door was opened to the possibility of irresponsible use of nuclear weapons by world powers.

Key Ideas

- There are three types of radiation: alpha particles, beta particles, and gamma rays.

- There are four types of nuclear reactions: radioactive decomposition, artificial transmutation, fission, and fusion.

- Radiation can be both beneficial and hazardous.

- Radioisotopes can be used in archeology, medicine, agriculture, energy, and forensic science.

These technicians are using remote manipulators to process cobalt-60, which will then be shipped to customers around the world for use in cancer treatment.

23.1
RADIOACTIVITY

Marie Curie (1867–1934)

Figure 23.2
Radioactive substances
must be labeled with this
radioactivity symbol.

The French physicist Henri Becquerel discovered radioactivity in 1896 while studying metals that fluoresced. He accidentally found that a covered photographic plate became exposed when it was placed near a piece of uranium ore. It was as though the ore produced something that had the same effect on film as exposure to light. The French scientists Marie and Pierre Curie later found that certain ores of uranium and radium give off invisible rays that cause this type of exposure. These ores are radioactive.

In 1909, the British physicist Ernest Rutherford discovered that the invisible rays from radium were actually composed of three different types of rays. When a beam of radioactive rays was passed near the pole of a magnet, the beam was split into three distinct parts. Some were positively charged particles, some were negatively charged, and others were uncharged rays. He named these alpha (α) particles, beta (β) particles, and gamma (γ) rays, respectively.

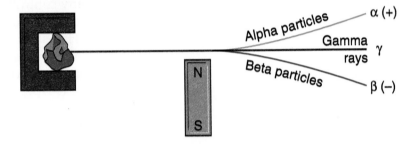

Figure 23.1   Splitting of a beam of radiation

The radiation emitted by radioactive substances can be dangerous if not handled properly. Such substances must be kept in specially shielded lead containers and labeled with the symbol shown in Figure 23.2.

23.2
TYPES OF
RADIATION

The three main types of radiation are alpha (α) particles, beta (β) particles, and gamma (γ) rays. They are emitted by radioactive nuclei and have energies far greater than those needed to break chemical bonds. When these high-energy particles and gamma rays pass through matter, they break up molecules. This can disrupt the normal operation of cells, and even kill them. Indeed, gamma rays are used routinely to destroy cancerous cells.

Alpha particles, emitted by only certain radioactive elements, are relatively large. Alpha particles travel at about one-tenth the speed of light, are positively charged and are not very penetrating. They can be stopped by about 7 cm of air or a sheet of paper. An alpha particle is composed of two protons and two neutrons (a helium nucleus).

As alpha particles travel through air, they collide with air molecules and knock the electrons belonging to the air molecules out of orbit. The air molecules then become ions. This type of radiation is called ionizing radiation.

Other radioactive atoms emit smaller particles called beta particles. A beta particle, which is an electron, travels much faster than an alpha particle. They are more penetrating than alpha particles, but beta particles can be stopped by thin sheets of metal or about 5 cm of wood.

Other radioactive atoms emit gamma rays. As the name suggests, gamma rays are not particles. They are similar to light and X rays but carry much more energy. Gamma rays are very penetrating—they can pass through 1 m of solid concrete and can be detected by photographic film.

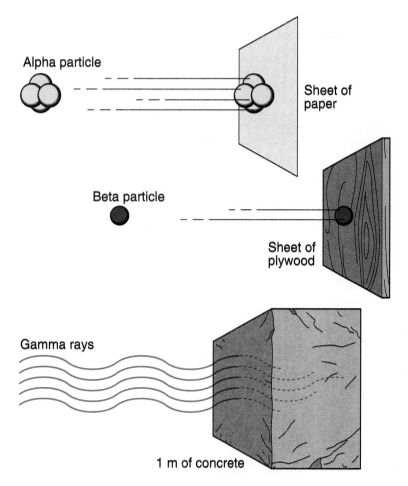

Figure 23.3   Different types of radiation have different penetrating abilities.

Table 23.1   Types of radiation

Type of Radiation	Composition	Symbol	Amount of Penetration
Alpha (α) particle	2 protons + 2 neutrons	$_2^4He$	a sheet of paper
Beta (β) particle	an electron	e^-	5 cm of wood
Gamma (γ) ray	similar to light waves	$_0^0\gamma$	1 m of concrete

Try These

1. Name the three main types of radiation.
2. Which type of radiation has
   a. the greatest mass?
   b. the least mass?
   c. the least penetrating power?
3. What is meant by the term "ionizing radiation"?

23.3 RADIOACTIVE DECOMPOSITION

There are four types of nuclear reactions: radioactive decomposition, artificial transmutation, fission, and fusion. We will discuss the first of these in this section and the remaining three in Section 23.6.

Only a few of the 90 elements that occur naturally in the earth's crust are unstable. The nuclei of these elements spontaneously emit alpha or beta particles to form new elements. When particles are given off in this manner and new elements formed, it is called radioactive decomposition. The two types of decomposition are called alpha decay and beta decay.

Alpha Decay

When a radioactive source undergoes alpha decay, an alpha particle is emitted from the nucleus. When this happens, the element loses mass. Uranium-238 is such an element. Its mass, 238 u, is due to the presence of 92 protons and 146 neutrons in its nucleus. When an alpha particle is given off, the nucleus loses 4 u in atomic mass and 2 in atomic number. The following equation shows radioactive uranium undergoing alpha decay. This reaction produces an alpha particle and a new, more stable element, thorium-234.

$$_{92}^{238}U \longrightarrow {}_2^4He + {}_{90}^{234}Th$$

Beta Decay

When a radioactive element undergoes beta decay, a beta particle is emitted from the nucleus. A beta particle is formed when a neutron in the nucleus of a radioactive element is converted into a proton and an electron. This causes the positive charge of the nucleus to increase by one. The thorium-234 nucleus formed from uranium-238 is an example of alpha decay. When it is formed, it is also unstable and decomposes to protactinium-234 by beta decay.

$$^{234}_{90}\text{Th} \longrightarrow {}^{234}_{91}\text{Pa} + {}^{0}_{-1}\text{e}$$

Try These

1. Name the four types of nuclear reactions.
2. Name the two types of radioactive decomposition.
3. What is the change in the atomic mass number and the atomic number during
   a. alpha decay?   b. beta decay?

23.4
WRITING NUCLEAR REACTION EQUATIONS

Alpha and beta decay are examples of nuclear reactions. Equations for these reactions are not difficult to write if the following rules are followed:

> 1. The masses on each side of the arrow must be equal.
> 2. The charges on each side of the arrow must be equal.

Example

Uranium also can be found as the isotope uranium-235. The equation for the alpha decay of this form of uranium, which has mass 235 u and atomic number 92, is

$$^{235}_{92}\text{U} \longrightarrow {}^{4}_{2}\text{He} + {}^{A}_{Z}\text{X}$$

Since the mass must be the same on both sides of the equation, the loss of 4 units of mass produces an element with a mass of 231 u:

$$^{235}_{92}\text{U} \longrightarrow {}^{4}_{2}\text{He} + {}^{231}_{Z}\text{X}$$

Charge must also be conserved. Since the alpha particle contains two protons, the nucleus of the new element must contain 90 protons:

$$^{235}_{92}\text{U} \longrightarrow {}^{4}_{2}\text{He} + {}^{231}_{90}\text{X}$$

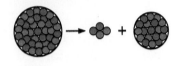

$$^{235}_{92}U \longrightarrow ^{4}_{2}He + ^{231}_{90}Th$$

Figure 23.4
Alpha decay of uranium

The new, more stable element can be identified by locating its atomic number in the periodic table. This element, atomic number 90, is called thorium:

$$^{235}_{92}U \longrightarrow ^{4}_{2}He + ^{231}_{90}Th$$

When the equation is balanced, the mass and charge on the left side of the equation equal the mass and charge on the right side of the equation.

Try These

1. Complete the following equations for alpha and beta decay reactions and identify element X:

   a. $^{238}_{92}U \longrightarrow ^{4}_{2}He + ^{A}_{Z}X$

   b. $^{32}_{15}P \longrightarrow ^{0}_{-1}e + ^{A}_{Z}X$

   c. $^{14}_{6}C \longrightarrow ^{4}_{2}He + ^{A}_{Z}X$

   d. $^{40}_{19}K \longrightarrow ^{4}_{2}He + ^{A}_{Z}X$

23.5   ACTIVITY
TRACKS IN A
CLOUD CHAMBER

We cannot see particles of radiation directly with our eyes. However, we can see the particles indirectly by using a cloud chamber. The cloud chamber contains moist air. Anything that moves through the chamber causes some of the moisture to condense from the air, leaving a track or vapor trail.

In this activity you will set up a cloud chamber and observe what happens when a radioactive source is placed in it.

Materials

radioactive source	cloud chamber
5 cm³ dry ice	dry wool cloth
2 mL alcohol	thermometer
eyedropper	damp cloth

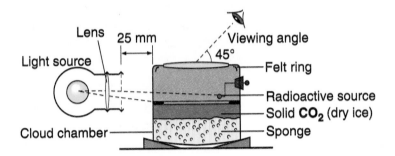

Caution: The radioactive source used in this activity is extremely weak. There is no danger if you use it carefully. Do not scrape it against any object or on your body. Wear safety glasses and gloves throughout this activity.

Method

1. Turn the cloud chamber over and unscrew the base.
2. Remove the sponge and cover the underside of the floor with 0.5 cm of dry ice.
3. Replace the sponge and screw on the base.
4. Turn the cloud chamber right side up.
5. Remove the lid and place the radioactive source in position on the inside of the chamber, as shown in the illustration.
6. Use the eyedropper to put about 2 mL of alcohol on the felt ring.
7. Clean the underside of the lid with a damp cloth and replace it.
8. Carefully level the cloud chamber by inserting the three wedges under it.
9. Charge the outside of the lid by rubbing it with a clean, dry, wool cloth. The charged lid will collect any charged particles in the chamber.
10. Direct a beam of light across the chamber as shown in the illustration.
11. Wait about five minutes. Observe the vapor on the inside of the chamber.
12. Note the temperature of the classroom. Look up the temperature of dry ice in a chemistry reference book.

Questions and Conclusions

1. What was the approximate air temperature in the classroom?
2. What was the temperature of the dry ice?
3. The alcohol vapor condensed somewhere between the cold bottom and the warm top of the cloud chamber. Describe the tracks that you saw in the cloud chamber.
4. Were all the tracks the same length or size?
5. What is the purpose of the alcohol?
6. a. What do you think caused the tracks in the cloud chamber?
   b. What was the origin of the tiny particles?

23.6
TRANSMUTATION, FISSION, AND FUSION

Artificial transmutation reactions are human-made and were first carried out by Ernest Rutherford. Today they are used to produce new elements with different atomic numbers. Element 105 was made by firing a nitrogen nucleus at the element californium, to form unnilpentium and four neutrons.

$$^{249}_{98}Cf + ^{15}_{7}N \longrightarrow ^{260}_{105}Unp + 4\ ^{1}_{0}n$$

californium + nitrogen $\longrightarrow$ unnilpentium + 4 neutrons

Fission reactions are the source of nuclear energy in both nuclear reactors and atomic bombs. Fission is caused by a neutron being fired at the nucleus of an atom. When this happens, the nucleus splits to produce new elements, additional neutrons, and a tremendous amount of energy.

$$^{235}_{92}U + ^{1}_{0}n \longrightarrow ^{141}_{56}Ba + ^{92}_{36}Kr + 3\ ^{1}_{0}n + energy$$

The **fusion** reaction is the underlying principle of the hydrogen bomb. When fusion occurs, large amounts of energy are given off as the nuclei of two different hydrogen isotopes fuse.

$$^{3}_{1}H + ^{2}_{1}H \longrightarrow ^{4}_{2}He + ^{1}_{0}n + energy$$

Try These

1. Write a word equation for the following reaction and name the element formed:

$$^{14}_{7}N + ^{4}_{2}He \longrightarrow ^{18}_{9}X$$

2. Balance the following nuclear equation and state the type of nuclear decay it undergoes:

$$^{230}_{90}Th \longrightarrow ^{A}_{Z}Ra + ^{4}_{2}He$$

23.7
THE HALF-LIFE OF RADIOACTIVE ISOTOPES

Radioactive elements decompose gradually over time. An element's **half-life** is the amount of time it takes for one half of the isotope to decompose into another more stable element.

For example, uranium-238 is an unstable isotope and will eventually decompose to a stable form of lead, **Pb**. It takes approximately 4.5×10^9 a (years) for one-half of uranium-238 to decompose into bismuth, a very long time indeed. Bismuth then decays to polonium and polonium to lead. This can be represented by the following equation:

$$^{238}_{92}U \xrightarrow{t_{1/2} = 4.5 \times 10^9 a} ^{210}_{83}Bi \xrightarrow{t_{1/2} = 5\ d} ^{210}_{84}Po \xrightarrow{t_{1/2} = 138\ d} ^{206}_{82}Pb$$

Table 23.2   The half-lives of some elements

Element	Half-life
I-129	1.7×10^7 a
I-133	21 h
Bi-210	5 d
Pb-214	26.8 min
Rn-222	3.8 d
Fr-222	15 min
Ac-227	22 a
Th-231	25 h
U-234	2.5×10^5 a

The number of radioactive atoms in a sample decreases to half the original number in each equal time period.

Figure 23.5   A typical radioactive decay curve

If 20 g of bismuth-210 was left for 5 d, only 10 g of **Bi-210** would remain. The rest would be converted to polonium-210. There would be a mixture of bismuth-210 and polonium-210. At the end of 10 d there would be one half of the remaining bismuth-210 left. The rest would be polonium-210.

Figure 23.6 is a concentration-versus-time curve for the decomposition of a radioactive element.

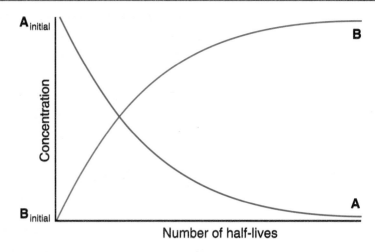

Figure 23.6   As **A** decomposes, the concentration of **B** increases. Note that **A** never reaches zero.

Example

Silver, $^{111}_{47}Ag$, decomposes by beta emission to form $^{111}_{48}Cd$.

1. Write the nuclear equation for the reaction.
2. If the half-life for silver-111 is 7.6 d, what fraction of the original amount of silver will be left after 76 days?

Solution

1. When silver decomposes, a neutron is changed to a proton and an electron and cadmium are the result.

$$^{111}_{47}Ag \longrightarrow ^{111}_{48}Cd + ^{0}_{-1}e$$

2. Find out how many half-lives have elapsed by dividing the half-life into the number of days elapsed:
   76 d/7.6 d = 10 half-lives
3. Use the number of half-lives found to calculate the fraction of the sample that remains. Since there are ten half-lives, multiply one half by itself ten times:

   After ten half-lives, 1/2048 of the original sample remains.

Try These

1. Define the term "half-life."
2. a. If you started out with 25 g of bismuth-210, what fraction will remain after 15 d?
   b. How much of the original 25 g is still bismuth-210 after 30 d?

3. You have 100 g of radioactive thorium-231.
    a. How long will it be before you have 50 g of this substance?
    b. How long will it be before you have 25 g?
4. a. How long does it take 10 g of **Fr-222** to become 2.5 g?
    b. What mass of francium-222 remains after 1 h?

23.8 ACTIVITY
FLIPPING COINS

Radioactive elements may undergo alpha decay, beta decay, or give off gamma rays. As elements emit alpha particles or beta particles, they eventually change into a substance that is not radioactive. In other words, the amount of radiation that can come from a given radioactive sample continually decreases. The time for one half of the mass of a substance to decay to another substance is called its half-life. The concept of half-life is illustrated in this activity.

Materials

100 pennies   graph paper
pencil

Method

1. Flip all 100 coins, one at a time.
2. Place the coins into two separate piles; one for all the coins that turned up heads, and one for tails. Record the number of heads and tails in a chart.
3. Place all of the pennies that turned up tails into a box, jar, or bag to separate them from the rest.
4. Flip all of the remaining coins again. Repeat step 2. Place all the tails in the container you are using for the coins.
5. Repeat step 4 until there is only one coin that turned up heads.
6. Graph the numbers from your chart.

Questions and Conclusions

1. What is a model?
2. In this model, what does the initial total number of coins represent?
3. What does the number of coins that turn up heads represent?
4. What do the tails represent?

23.9 ACTIVITY
CARBON-14
DATING

Carbon-14 is a radioactive element that is produced high in our atmosphere and becomes part of the CO_2 in the air.

Plants take in CO_2 which contains **C-14**. Humans and animals eat these plants. When these organisms die, the intake of CO_2 and **C-14** stops. The carbon-14 present at that moment begins to decay. A procedure called carbon-14 dating can be used to give a good estimate of the age of the organism.

The late Dr. Willard F. Libby, then a professor at UCLA, won the Nobel Prize for chemistry in 1960 for developing the technique of radiocarbon dating. If an organism has been dead for a short time, it contains a relatively large amount of carbon-14. If the organism has been dead for a long time, it contains a small amount of carbon-14.

Radiocarbon dating

In this activity you will sketch the half-life curve for **C-14** decay, using the information provided in the following chart.

Counts per min per gram C	Number of Years
7.00	5 700
3.75	11 400
2.00	17 100
1.00	22 800

Method

1. Draw a half-life curve for carbon-14 decay from the data given in the chart.

Questions and Conclusions

1. How old is a dead tree that emits 7 counts per minute per gram of carbon?
2. How old is a human skeleton that emits 2 counts per minute per gram of carbon?
3. How old is a skeleton that emits 3.75 counts per minute per gram of carbon?
4. How many counts per minute per gram are emitted by a specimen 11 400 a old?
5. How many counts per minute per gram are emitted by a specimen 22 800 a old?
6. In what range of years does the carbon-14 method begin to fail?

23.10 RADIOACTIVITY IN MEDICINE

Radioisotopes

Much of what we know about biology depends on our ability to trace the path of a substance through a biological system. A common way of tracing substances is by the use of radioactive tracer isotopes, or radioisotopes. A radioisotope is a radioactive form of an element, including common biological elements, such as potassium. It is now possible for researchers and medical professionals to purchase many kinds of compounds which are labeled with radioactive isotopes.

The discovery of radioactivity has helped to solve many biological and medical problems. Radioactive elements are now being used as research tools in medicine, biology, and biochemistry. They are used to detect stomach ulcers, brain tumors, and lesions in blood vessels. Plant scientists use radioactive elements to trace the movement of substances in plants. Isotopes used in this way are called tracer isotopes.

When a circulation problem is suspected, the use of radioactive isotopes as tracers plays an important role in monitoring the course of blood through the body. This is accomplished by injecting serum containing sodium-24 into the bloodstream. This isotope is used to replace naturally occurring sodium-23 in the blood.

A Geiger counter is placed near the area where the circulation trouble is suspected. The Geiger counter will indicate when the radioactive sodium-24 reaches the trouble spot, since the counter detects radioactivity. By using a stopwatch, the

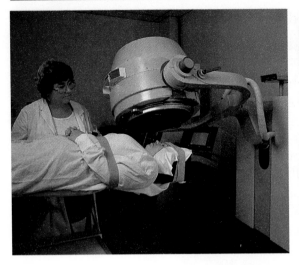

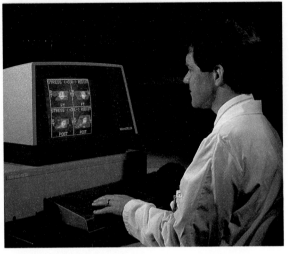

Tracer isotopes are used to create images that help indicate the size, shape, and location of tumors.

rate of blood circulation in any part of the body can be timed. This technique can be used to help doctors determine, for example, whether an injured limb should be amputated.

Iodine-131

Iodine-131 is another isotope used in medicine. It decays to form xenon-131 according to the following equation:

$$^{131}_{53}\text{I} \longrightarrow {}^{131}_{54}\text{Xe} + {}^{0}_{-1}\text{e}$$

Iodine-131 is used as a diagnostic tool in cardiovascular and thyroid malfunctions. In the case of suspected heart malfunctions, serum containing iodine-131 is injected into the patient's bloodstream and a counter is placed on the patient's chest.

The counter traces the route of the iodine-131 through the chest. An area that does not give off beta emissions indicates an area of the heart that is not receiving an adequate supply of blood. An inadequate supply of blood to any area of the heart is one of the major causes of heart attacks.

The thyroid gland is a part of the body that requires iodine to function properly. When iodine-131 is ingested in the form of sodium iodide in solution, it should accumulate in the thyroid gland. If it does not, the thyroid is malfunctioning.

Iodine-131 is also used to treat thyroid cancer. In this treatment, large amounts of iodine-131 are given to the patient to kill the cancer cells.

Gamma radiation is used to sterilize surgical instruments and medical products.

Table 23.3   Radioactive isotopes used in medicine

Isotope	Use
fluorine-18	bone imaging
gallium-67	imaging for Hodgkin's disease
rubidium-81	myocardial imaging
xenon-131	lung ventilation
ytterbium-169	spine and brain imaging

Try These

1. Why is the radioisotope iodine-131 used in diagnosing disease?
2. How is sodium-24 used in medicine?
3. Give one reason why an area of the heart might not be receiving blood.
4. What substance do red blood cells carry that heart muscle cells need?

Cobalt Radiation Therapy

Radiation from cobalt-60 can be used to kill cancer cells. This technique has saved millions of lives.

Cobalt radiation therapy, or radiotherapy, beams radiation into the body to destroy cancerous tumors. Radiotherapy was used as early as the 1800s.

Close to half a million patients annually are treated for cancer in over 70 different countries, using cobalt therapy. Cancer in many parts of the body may be treated with a beam of radiation produced by a "teletherapy" machine containing several thousand curies (a unit of radiation) of **Co-60**. The patient is placed inside a shielded room, and the patient is shielded, so that only the affected site is treated. A team of specialists then administers the radiation to the targeted area, destroying the tumor.

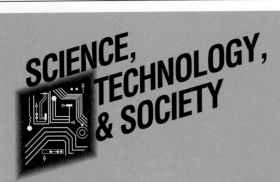

Food Irradiation

The foods we eat may have bacterial infections or animal infestations. These could cause the food to spoil, make us sick, or even cause death. More than one-quarter of the world's food production goes to waste because it is destroyed by bacteria, insects, or rodents. Because of the need to avoid food spoilage and protect human health, the food industry and government scientists are continually searching for effective methods of preserving food while keeping it wholesome for us to eat.

Food irradiation is the most recently developed method for food preservation. Salting and drying are methods which are more than one thousand years old, while pickling and freezing have been in use for about 175 years.

Food irradiation is a non-chemical process which preserves food, making it longer-lasting and safer to eat. Many foods can be irradiated without significant change in taste or nutritional value. Irradiation has several beneficial effects. First, it kills the micro-organisms that normally cause food to spoil. Second, irradiation kills micro-organisms and parasites that can cause human illness. Third, irradiation can slow the ripening process of certain fruits and vegetables, and can prevent the sprouting of root vegetables like potatoes and onions.

How is Food Irradiated?

Food is irradiated in a way that is similar to the way luggage is X-rayed at airports. Food is passed through a thick-walled chamber on a conveyer belt. The chamber contains a source of radiation. The ionizing radiation (gamma rays) from the radioactive source passes through the food, destroying insects, bacteria, and other micro-organisms. Because of the nature of radiation, the food does not have to be removed from its crates or boxes to be treated. Food irradiation does not make the food, or the consumer, radioactive!

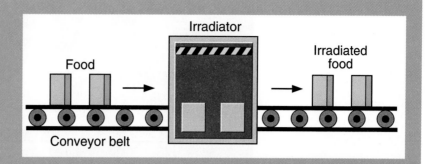

Food irradiation is often done using cobalt-60, an artificial isotope which also has many medical and industrial uses. Cobalt-60 has also made its mark in agriculture. Radiation from this isotope can change the genetic structure of plants and hence their behavior. Two examples of this are faster growing rates and increased crop yields. For some crops, a faster growing rate allows more than one crop to be produced per growing season. In addition, some cereal crops have been found not only to mature faster, but also to produce higher yields after bombardments with radiation.

23.11 ACTIVITY
THE EFFECT
OF RADIATION
ON PLANTS

The high-speed particles given off by radioactive substances can affect living things. To illustrate this, you will compare the germination of normal seeds with seeds that have been exposed to various amounts of radiation.

Materials

package of irradiated seeds
package of non-irradiated seeds
planting containers
labels

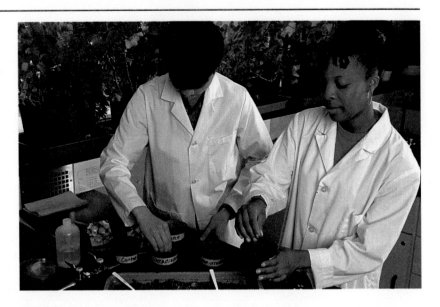

Method

1. Draw a chart to record your data. Include the headings Radiation Dose, Number of Seeds Germinated, and Percentage of Seeds Germinated.
2. Open the package of seeds and read the instructions.
3. You will find seeds that have been exposed to varying amounts of radiation, given in grays or rads. The gray (Gy) is the SI unit for radiation; the rad is a non-SI unit. Sort the seeds into separate piles according to the amount of radiation they have received.
4. Plant the seeds according to the instructions on the seed packages. Keep the seeds separate according to the amount of radiation received.
5. Plant a few non-irradiated seeds as controls. Make sure that all of the seeds receive the same amounts of light, water, and so on.
6. Place the seeds in a safe environment and allow them to grow for several weeks.
7. Observe the containers daily. Complete your chart.
8. Obtain a class set of data by adding together the number of seeds in each category.

Questions and Conclusions

1. Which amount of radiation most affected the germination of the seeds?
2. Did radiation increase or decrease the germination rate?
3. Did you obtain any plants that were quite different than the others? If so, how do you account for the difference?
4. According to the results of this activity, what level of radiation is harmful to plants?

23.12
RADIOISOTOPES IN FORENSIC SCIENCE

Napoleon Bonaparte
(1769–1821)

Many forensic techniques cannot detect chemicals that are present in concentrations of less than one part per million. However, the use of neutron activation analysis has eliminated this problem.

When a neutron hits an element, the radiation characteristic of that element is given off and the element can be identified even if it is present in concentrations of less than one part per million. The sample being tested does not have to be pretreated and it is not destroyed in the test. The following story about Napoleon Bonaparte is a good example of how useful this technique can be.

Napoleon Bonaparte lived in exile on St. Helena Island for the last years of his life. Before he died, he complained of excruciating pain in his stomach and was certain that his jailers were poisoning him.

Fortunately, two of his valets saved a lock of his hair. When these hair samples were bombarded by neutrons more than a century later in a laboratory at the University of Toronto, high levels of arsenic were found. Napoleon was right. He *was* being poisoned.

Neutron activation analysis has many other applications. The authenticity of paintings can be determined by analyzing the mineral content of the paint used.

The results of neutron activation analysis in criminal investigations are often dramatic. The basic idea is to match the elements in paint, soil, cosmetics, and so on found at the scene of a crime with those found on a suspect. This technique can use a wiping taken from a suspect's hand and can determine not only whether the person has fired a gun recently, but also the type of ammunition used.

Try These

1. Why is neutron activation analysis a good forensic technique?
2. What killed Napoleon?
3. How could traces of arsenic have been present in Napoleon's hair samples?

OPPORTUNITY

Radiation Therapy Technician

Large hospitals and medical centers do extensive work in nuclear medicine. Through the use of radioactive materials, doctors, nurses, and radiation therapy technicians can diagnose and treat patients for cancer and other diseases.

Radiation therapy technicians treat cancer patients with intense doses of radiation. The same kind of treatment is given to kidney transplant patients. In consultation with the doctor, technicians can calculate the radiation dose, prepare the patients, and administer treatment.

Radiation therapy technicians are able to develop close relationships with patients because they usually see these patients five days a week during the treatment period of six to eight weeks.

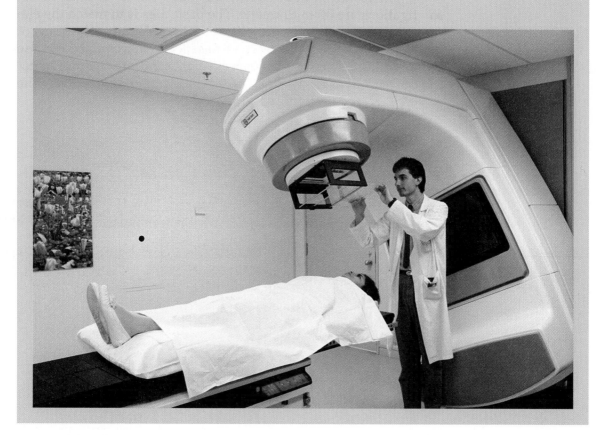

Exercises

Review

1. a. When was radioactivity discovered and by whom?
   b. How was this discovery made?
2. a. Name three types of radiation.
   b. Who discovered and named them?
3. Compare the composition, mass, and penetrating power of the three types of radiation.
4. Name four types of nuclear reactions.
5. What two properties must be conserved in a nuclear reaction?
6. Differentiate between fission and fusion.
7. Define the term "half-life."
8. A radioactive sample has a half-life of 7 h. How much of it will remain after 28 h?
9. State two rules for writing nuclear reactions.
10. Identify the new element formed in the following equation:

$$^{218}_{84}\text{Po} \longrightarrow {}^{4}_{2}\text{He} + {}^{A}_{Z}\text{X}$$

11. What do tracks in a cloud chamber indicate?
12. List and briefly describe some of the benefits we get from radiation.
13. Briefly explain how neutron activation analysis works.
14. Give two examples of uses of neutron activation analysis.

Extension

1. Radioactive elements are often used in scientific and medical research. How are these elements used? What precautions must be taken when working with radioactive substances?
2. Find out how a Geiger counter works and describe its uses.
3. Use your library resource center to find out the cause of Marie Curie's death.
4. Tritium is an isotope of hydrogen and has the symbol **T**. It is produced in the atmosphere by cosmic rays. It falls to the ground as T_2O. Tritium has a half-life of 12.5 a. How could you use this information to date vintage wines?
5. Suggest an experiment to synthesize element 108.
6. Research the difference between radiation therapy and chemotherapy. Why are they used for different types of cancer?

CHAPTER 24
NUCLEAR ENERGY

Fission and fusion reactions can produce vast amounts of energy. Modern technology has enabled us to take large amounts of usable energy from nuclear reactions in a controlled way. This source of energy is clean, safe, and relatively inexpensive.

However, many controversial issues have arisen. Nuclear reactors produce waste which must eventually be disposed of. There have been some nuclear accidents resulting in leaks of radioactive materials. These materials are harmful to living things and the environment. Some people feel that the risk of accidents is too great compared with the benefits to society from nuclear energy.

Today, we have more information about both the benefits and hazards of nuclear energy. As our understanding of atomic theory grows and new technologies are developed, the problems associated with nuclear energy will continue to decrease.

Key Ideas

- Modern atomic theory and the discovery of radioactivity led to the use of nuclear energy.

- There are several different types of nuclear reactors.

- Uranium, the fuel in nuclear reactors, undergoes nuclear fission.

- The nuclear waste that results from nuclear reactors poses a waste disposal problem.

- Because there are potential hazards associated with it, nuclear energy is often controversial.

A heavy water plant. Heavy water is used in nuclear reactors.

24.1
NUCLEAR ENERGY

Demands for electricity
are likely to increase
in the future.

A century ago, coal and wood were both important sources of energy for the production of heat and electricity. However, the use of these fuels has decreased over the past 50 years while the use of oil has increased.

Canada and the United States are fortunate to have a natural abundance of energy sources such as oil, natural gas, coal, and uranium. However, the use of fossil fuels has led to many concerns about future supplies, costs, and pollution.

At the same time, there has been a constantly increasing demand for more electricity at a reasonable cost and this will continue to be the situation in the future. Some people feel that the least expensive, most efficient, and least environmentally hazardous way to generate this electricity is by the fissioning of radioactive materials.

Many people associate increased amounts of electricity with increased numbers of nuclear power plants and increased pollution of the environment. However, most people are unaware that natural radioactivity has always existed and is already in our environment from many other sources aside from nuclear energy.

Uranium and thorium are two examples of naturally occurring radioactive elements found in the earth's crust. They can and do undergo natural fission reactions if the proper conditions exist.

A nuclear power plant

Try These

1. Why does the demand for electricity increase?
2. What problems are associated with using fossil fuels to create electricity?
3. Use the library resource center to write a short report on a naturally occurring prehistoric nuclear reactor.

24.2
URANIUM—THE NUCLEAR FUEL

As a result of the U.S. demand for uranium during and after World War II, uranium mining grew as an industry in Canada. After the war, Canada made use of this industry to begin atomic energy research. Today, Canada and the United States are both world leaders in nuclear research.

Natural uranium is made up of two isotopes, uranium-235 and uranium-238. In a typical sample of uranium, 99.3% is uranium-238, which is not fissionable, and 0.7% is **U-235,** which is fissionable.

The uranium pellets used as fuel exist as uranium oxide **(UO_2),** which is prepared from uranium ore, **U_3O_8.** Uranium oxide pellets are yellow, hence their nickname, yellow cake.

Uranium undergoes fission (splitting) to produce energy in a nuclear reactor. When fission occurs, a heavy nucleus such as uranium-235 splits into two smaller nuclei. The decomposition rate of uranium-235 can be increased by bombarding its nucleus with neutrons. When a neutron is fired at a uranium-235 nucleus, it splits the nucleus into two smaller nuclei and three neutrons.

$$^{235}_{92}U + ^{1}_{0}n \longrightarrow ^{141}_{56}Ba + ^{92}_{36}Kr + 3\,^{1}_{0}n + energy$$

The two nuclei formed are barium and krypton. The neutrons produced in the reaction can cause additional reactions, which in turn produce more neutrons. This is why fission reactions are called chain reactions.

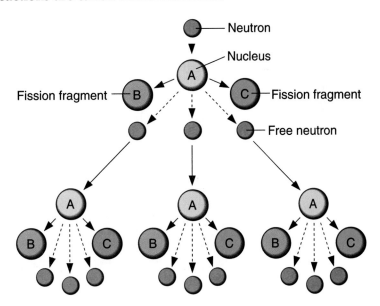

Figure 24.1   A chain reaction

Try These

1. What is yellow cake?
2. How does uranium decompose?

24.3 ACTIVITY
THE YELLOW CAKE CAPER

It was 8:00 a.m. when the phone rang in Analytical Al's lab. Uranium Refinery B was calling in a panic over the disappearance of ten barrels of yellow cake, valued at $ 400 000. The police had tracked down two suspects and what they think may be the stolen yellow cake.

Al was asked to find out whether the recovered yellow cake was from Refinery B. The researcher-turned-detective hoped to identify the yellow cake by trace-metal analysis. Al obtained samples from the surrounding refineries and examined each refinery's processing methods.

He reasoned that there would be two possible sources of trace metals in the samples. One source could be linked with the location of the ore formation from which this particular yellow cake had been mined. Different locations contain different types of trace metals. The second source would be the process used to precipitate the yellow cake. Different refinery locations have different trace metals in their soils and water. Al came up with the following information:

Refinery	Type of Formation Mined	Purification Process
A	nickel and manganese	precipitation by ammonium hydroxide
B	iron salts	precipitation by sodium hydroxide

Like Analytical Al, you do not have time to wait for a tip-off to the crime. Speed is essential, so you will use flame tests to identify the different metals present in each sample.

Materials

burner
inoculating loop
samples (4)

Caution: Wear goggles while doing this activity.

Method

1. Place one of the samples on an inoculating loop and hold it in the flame of the burner.
2. Record the color of the flame.
3. Repeat steps 1 and 2 for each of the other three samples.

Questions and Conclusions

1. What trace metal was found in each of the samples?
2. Is the recovered yellow cake from Refinery B?
   What evidence do you have?
3. Is this sufficient evidence to prosecute the suspects?
4. What other tests could be carried out to verify your results?

24.4

NUCLEAR REACTORS

A nuclear reactor is a device that converts nuclear energy into heat energy. In nuclear power plants, this heat is then used to make steam to drive turbines that generate electricity. The reactor is made up of a fuel, a moderator, and a coolant. The coolant keeps the system from overheating. The fuel in commercial reactors is usually uranium-235.

The neutrons must be traveling at slow speeds to cause **U-235** to undergo fission. This is achieved by a moderator. The type of moderator depends on the type of nuclear reactor and the fuel used. Some common moderators are water and carbon.

Pressurized Water Reactor (PWR)

The fuel used in pressurized water reactors is uranium oxide enriched with uranium-235. This uranium is placed in tubes of zirconium alloy 3.5 m long, which form single fuel elements. About two hundred of these fuel elements make up the core of the reactor. The core is contained in a 20 cm-thick pressure vessel.

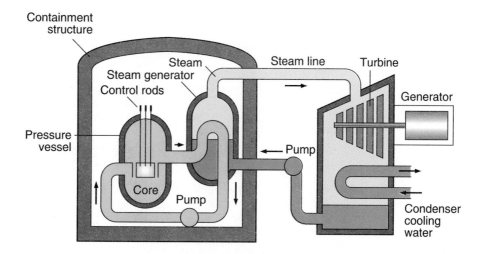

Figure 24.2   A pressurized water reactor

The reaction rate is increased or decreased by inserting or removing control rods made of boron or cadmium steel. The ideal reaction rate occurs when the number of neutrons produced equals the number of neutrons absorbed. If the ratio falls below one, fission stops. If the ratio rises above one, fission increases.

Nuclear power operator

Boiling Water Reactor (BWR)

The only difference between the boiling water reactor and the pressurized water reactor is that the cooling water is boiled in the reactor core of the boiling water reactor. The steam generated goes directly to the turbines to produce electricity.

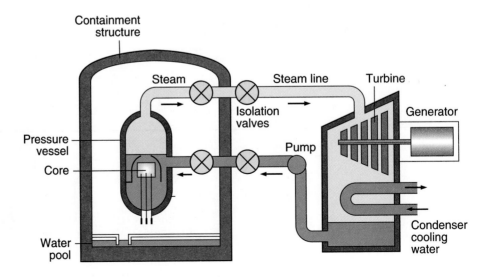

Figure 24.3   A boiling water reactor

24.5
THE CANDU REACTOR

The CANDU (**Can**ada **d**euterium **u**ranium) reactor is a pressurized heavy water reactor. As the name indicates, it is a Canadian invention. Heavy water is the moderator, and uranium is the fuel.

Deuterium is an isotope of hydrogen that contains one neutron in its nucleus instead of none. For this reason, water containing deuterium is called heavy water; its formula is D_2O.

The reaction vessel in the CANDU reactor is cylindrical, lies on its side, and uses heavy water as its moderator. Natural uranium is used in the fuel rods. The fuel is placed in the part of the reactor called the calandria (the core) and when fission transforms the uranium into its fission products, heat is given off.

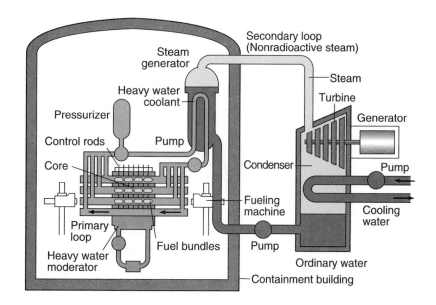

Figure 24.4   A CANDU reactor

Heat from the reaction warms up the heavy water that flows into the heat exchanger. This causes the water to boil, changing it into steam that causes the turbines to rotate and electricity to be generated.

To slow down the reaction in a light water reactor, the level of light water (ordinary water) is increased. It absorbs the fast moving neutrons so that they are not available for fission.

To speed up the reaction in a heavy water reactor, the level of heavy water is increased. The heavy water moderator slows down the neutrons so that more fission reactions can occur.

Figure 24.5 on page 440 shows how the rate of a nuclear reaction can be increased or decreased using light or heavy water.

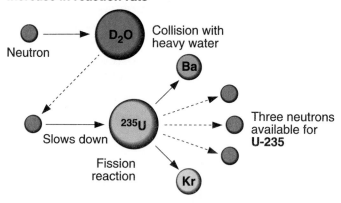

Figure 24.5   The process of controlling the fission reaction

Try These

1. Describe how a moderator works in fission reactions.
2. What is meant by a fission reaction?
3. Describe a chain reaction.
4. Why is water containing deuterium called heavy water?
5. Describe how the rate of a fission reaction can be controlled.

24.6
THE NUCLEAR ISSUE

Because electricity produced by nuclear reactors is the least expensive and most efficient method, the number of nuclear generating stations is likely to increase in the future. This causes many people to be concerned about radioactive material being released into the environment and injuring residents in nearby communities. It is helpful to understand two important factors about radioactivity and nuclear energy.

1. A nuclear explosion cannot occur at a nuclear generating plant. A nuclear explosion requires highly enriched uranium or plutonium and a triggering device, neither of which is present in nuclear power plants.
2. Nuclear energy contributes less radioactivity to the environment than most other sources of radiation.

The radiation present in our environment today comes from natural and human-made sources. The natural sources include cosmic rays, the sun, and deposits of uranium, radium, and thorium in the earth's crust. The human-made sources include X rays and radioactive materials used in medical diagnosis and therapy, mining minerals, burning fuels, fallout from past testing of nuclear weapons, consumer products such as color televisions, smoke detectors, luminous dial watches, and the commercial operation of nuclear power plants.

Table 24.1   Average annual human radiation exposure from some common sources

Source	Dose (μSv/year)
Naturally Occurring Radiation	
cosmic rays	350
terrestrial radiation (rocks, buildings)	500
radioisotopes ($^{40}_{19}$K and $^{226}_{88}$Ra) in bones	300
radon in the air	250
Subtotal	1400
Artificial Radiation	
medical sources	
diagnostic X rays	500
therapeutic and radioisotopes	70
fallout from nuclear weapons tests	20
nuclear power plants	10
Subtotal	600
Total	2000

Radiation is measured in sieverts (Sv), the SI unit for the dose equivalent of ionizing radiation equal to one joule per kilogram. It is estimated that each individual receives an annual dose of 1990 μSv from all the sources mentioned above except nuclear energy. When the dose of radiation from nuclear energy is added to the annual individual dose, the total increases to 2000 μSv.

Studies over several decades indicate that, below a level of 100 000 μSv, the effects of radiation on living organisms are negligible. Above that level, cancer and genetic defects may occur.

Today's society has limited technical, material, and financial resources. Therefore, the risks of one course of action must be carefully evaluated against the risks of another course of action. If this is not done, large sums of money may be spent to reduce a small risk even further, while greater hazards may receive little attention.

SCIENCE & TECHNOLOGY

Is There Anyone Else Out There?

For thousands of years, people have watched the heavens. Even a casual glance at the night sky reveals a universe of unfathomable size and incredible beauty. The constellations, like the Big Dipper and Orion, and the ancient myths and beliefs that surround them, have fascinated people for centuries.

The invention of the telescope, modern research satellites, and the advent of manned space travel have produced a wealth of information about our universe. Today, powerful telescopes scanning the electromagnetic spectrum reveal a cosmos of intricate structure and incredible violence. The universe is a "battleground of forces" that shape matter and release huge quantities of energy. This energy reaches the earth in the form of light and other kinds of radiation and is detected by huge radiotelescopes.

These radio telescope antennas are part of the Very Large Array in New Mexico.

Researching the Origins of the Universe

Many astrophysicists believe that the universe came into being about fifteen billion years ago as the result of a huge explosion—the "Big Bang theory." This explosion sent immense quantities of matter and energy traveling billions of kilometers in all directions. American scientists have discovered the existence of radiation patterns in space that may mark the beginning of time itself. The remnants of the great blast, in the form of microwaves known as cosmic background radiation, were detected in 1964. Cosmologists have designed dozens of experiments to help them analyze this radiation. To a cosmologist, cosmic background radiation represents the earliest information available about the formation of the universe. Sophisticated computers are used to constantly process the radiowave signals.

Other forms of research are used to speculate about the origins of the universe. For example, in the last century, astronomers discovered that there are billions of galaxies, each with billions of stars like our sun. Some cosmologists are trying to unravel the mysteries of the universe by studying the size, location, and distribution of the galaxies. So far, cosmologists have photographed and mapped three million galaxies.

Galaxies form clusters of galaxies; clusters are organized into super-clusters and finally super-cluster complexes. What explains this remarkable cosmic architecture? If the matter scattered throughout the universe by the Big Bang was scattered evenly, as cosmologists believe, how were solar systems, galaxies, clusters of galaxies, and super-clusters formed? The data received by radiotelescopes in the form of electromagnetic radiation may provide the answer. Physicists continue to analyze many years worth of data, and that material could provide further insights into the beginnings of space and time.

The Orion nebula

24.7
STATISTICS
AND RISK

Statistics can be used to help make decisions. Statistics are numbers that are calculated using mathematical models. On the basis of these numbers, both probability and risk concerning an event can be calculated. Probability is the mathematical likelihood that a certain event will happen. Risk is the degree of chance associated with that probability. An event that has a high probability of happening, or is very likely to happen, corresponds to a situation of high risk. An event that has a low probability of happening corresponds to a situation of low risk. The following chart shows the estimated probability of death resulting from various types of accidents.

Type of Accident	Number of Fatal Injuries
automobile	1 in 4000
drowning	1 in 30 000
airline	1 in 100 000
lightning	1 in 2 000 000
nuclear	1 in 5 000 000 000

Try These

1. Which type of accident has the greatest probability of occurring?
2. Which type of accident has the least probability of occurring?
3. According to these statistics, which type of accident presents the greater risk, automobile or nuclear?

24.8
MELTDOWN

In the history of nuclear power technology, serious accidents have been very rare. A complete meltdown has never occurred.

In the case of a complete meltdown, the reactor would lose all of its coolant, which would cause the core to heat up uncontrollably. Eventually the intense heat from the core could melt the reactor's steel and concrete housing. Radiation could be released into the immediate environment. However, nuclear experts claim that only 1 in 5000 such meltdowns would release enough radiation to cause fatalities.

The 1979 partial meltdown at Three Mile Island in Pennsylvania was not as disastrous as it could have been had there been no containment building present. The building

The Chernobyl reactor
in the Ukraine

prevented the radioactive material from escaping. Very little radiation was released and there was little danger to the general public.

In April 1986, the water system of the Chernobyl reactor in the Ukraine (then part of the U.S.S.R.) failed. The power plant overheated and a partial meltdown occurred. The Chernobyl reactor had no containment building. As a result, a series of events occurred that resulted in the worst nuclear accident to date.

Try These

1. Describe a meltdown.
2. Do you believe that the production of nuclear energy should be halted because of the possibility of nuclear accidents? Give reasons for your answer.

24.9

TECHNOLOGY AND NUCLEAR WASTE

Like many other industries, the nuclear industry has a waste disposal problem. However, its problem is unique because the waste is radioactive and requires special handling and disposal methods.

Storage of nuclear waste has been the solution to date. This method seems fine for low-level radioactive waste, but it is not the answer for high-level radioactive waste.

High-level radioactive waste takes hundreds of years to decay. In the process it gives off a great deal of heat. Presently, all forms of high-level wastes are stored underground in tanks that consist of steel liners encased in concrete.

Research is being done on the storage of radioactive wastes in natural geological formations. There are two suitable formations: salt deposits and bedrock storage. Both, however, must be well removed from populated areas and clear of the groundwater table.

Salt Deposits

Salt is a good storage medium because it is not associated with drinking water. It is strong enough to support a mine without collapsing. The salt can also change shape to seal a fracture. Waste could be stored in these mines until the radioactive decay process is complete.

Bedrock Storage

Bedrock storage can be used to store liquid radioactive waste. This type of storage uses underground tunnels cut into the solid bedrock with a clay overlay. A clay overlay is best for this type of storage because it is impermeable to liquids and seepage to the outside cannot occur.

Try These

1. What makes the problem of waste unique in the nuclear industry?
2. List the advantages and disadvantages of storage as a means of radioactive waste disposal.
3. What region of the country would you select as a storage site for radioactive waste? Explain why.

24.10 ACTIVITY
THE PROCESSING OF LOW-LEVEL NUCLEAR WASTE

Low-level radioactive waste accumulates in large volumes that must be disposed of at frequent intervals. Most of the low-level waste produced during the whole history of nuclear power is being stored in above-ground containers. The process used to decrease the residue is based on separation methods with which you are familiar.

Under alkaline conditions, radioactive particles will react with a coagulant. A coagulant is a substance that causes a liquid to clot or form a jelly-like material. This coagulated mixture is then separated and packed for disposal. It contains most of the fission products. The water used for this process is pumped through ion-exchange columns where the remaining waste is removed.

In this activity you will be using ion exchange to separate a mixture.

Materials

30 cm glass column
one-holed rubber stopper
retort stand
utility clamp
glass wool
medicine droppers (2)
10 cm glass tube
5 cm rubber tube sleeve with glass bead
distilled water
sodium chloride solution
cation exchange resin
calcium chloride solution
sodium carbonate solution

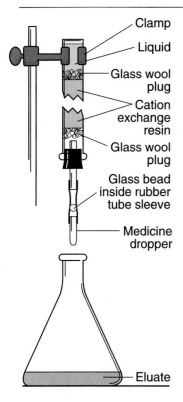

Clamp
Liquid
Glass wool plug
Cation exchange resin
Glass wool plug
Glass bead inside rubber tube sleeve
Medicine dropper
Eluate

Method

1. Set up the ion exchange column as shown in the margin.
2. Add distilled water to the column, which contains the cation exchange resin.
3. Squeeze the glass bead in the rubber tube sleeve until the level of water has dropped down to the upper glass wool plug.
4. Pour 10 mL of calcium chloride solution onto the column.
5. Squeeze the rubber tube sleeve again until the water level is equal to the level of resin in the column.
6. Add 30 mL distilled water and allow it to run through the resin until the level is at the upper glass wool plug.
7. Discard the solution that has passed through the resin. This solution is called the **eluate**.
8. Run 10 mL of sodium chloride solution through the column.
9. Flush the column with 10 mL of distilled water.
10. Test the eluate by adding six drops of sodium carbonate solution. Record your results.
11. Test the original calcium chloride solution by adding six drops of sodium carbonate solution.

Questions and Conclusions

1. For what ion is sodium carbonate a test?
2. What ion exchange occurred in step 8?
3. How is ion exchange applied in the processing of low-level radioactive waste?

Postlab Discussion

Figure 24.6 illustrates how ions are exchanged between an ion exchange column and a solution. As a solution of sodium ions is poured over the column, the calcium ions and sodium ions trade places. The sodium ions become attached to the column, and the calcium ions are released into the solution. A test to prove calcium ions are present in the eluate is to add sodium carbonate. When it is added to the eluate, the following precipitation reaction occurs:

$$Ca^{2+} + 2\ Cl^- + 2\ Na^+ + CO_3^{2-} \longrightarrow CaCO_3(s) + Na^+ + Cl^-$$

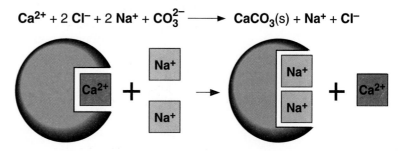

Figure 24.6
Exchange of ions in a column

24.11
PLUTONIUM

One of the byproducts of fission reactions is the element plutonium. Although plutonium is found in very small quantities in nature, the vast majority of it is artificially produced in nuclear reactors.

Plutonium is a metal, more dense than uranium, with all of the characteristics of a metal. However, all of its isotopes are radioactive, making it a rather unusual element. Plutonium has a half-life of 24 390 a and decays by alpha particle emission. Because of this, it is extremely toxic, but it is not a hazard to human beings as long as it remains outside the body. The only way it can reach the interior of the body is through inhalation, ingestion, or through cuts. When it does this, it can damage sensitive internal tissues.

Plutonium results from the collision of one of the three neutrons produced in the normal fission reaction with a highly energetic atom of uranium-238. This reaction produces **U-239**, an unstable isotope that quickly decays to plutonium-239.

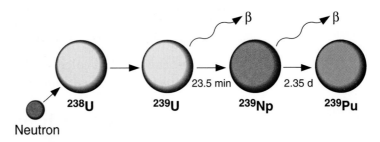

Figure 24.7   The formation of plutonium-239

About half of this plutonium escapes fission and remains, together with the used uranium-235, in the fuel rods removed from the reactor. This plutonium could be used as a fissionable material in nuclear reactors. Recycling these rods and using the still fissionable material contained in them seems much more logical and economical than storing them underground for many years. This would greatly reduce the amount of uranium ore required as fuel, further decrease environmental threats, and solve future storage problems.

Because plutonium has such a long half-life, it is used to produce electricity in thermoelectric generators, cardiac pacemakers, satellites, navigational beacons, and remote weather stations.

If it is combined with certain other metals in compounds, it emits a constant number of neutrons over a long period of time. This makes it ideal for use in reactor start-up and instrument calibration.

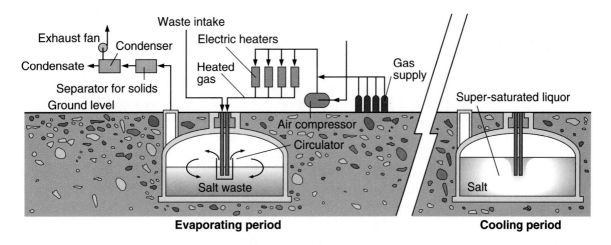

Evaporating period   **Cooling period**

Figure 24.8   A proposed storage facility for plutonium

Try These

1. Describe the formation of plutonium-239.
2. Give two reasons why plutonium is an important element.

24.12 ACTIVITY
DEBATE THE NUCLEAR ISSUE

Human activities have had long-term effects on our environment. We now have a much better ability to predict the long-term effects of our actions. Energy production is a primary source of pollution and other environmental problems. While conservation can reduce energy consumption, we still need large amounts of energy. There is a great deal of controversy over what is the safest, cleanest, and least expensive method of power generation.

In this activity you will debate the best method of generating electricity. While the two main options are nuclear and fossil fuels, some debaters could argue for other options such as hydroelectricity. Among the factors that should be considered are safety, cost, environmental effects, disposal of wastes, supply of fuel, and available technology. Your teacher will organize the format of the debate.

Geiger Counter Technician

A technician working in this field could be involved in exploration for uranium. A Geiger counter is an instrument used to detect levels of radiation. As a Geiger counter technician working in uranium exploration, you would stake out an area of land, and then cover the entire area taking Geiger counter readings. If the levels of radiation were sufficiently high, the company would consider mining the area.

Aside from exploration, Geiger counters are used extensively for health and safety checks by Geiger counter technicians. Industries which use or produce radioactive materials and nuclear power plants must monitor the levels of radiation in the workplace to ensure the safety of their employees.

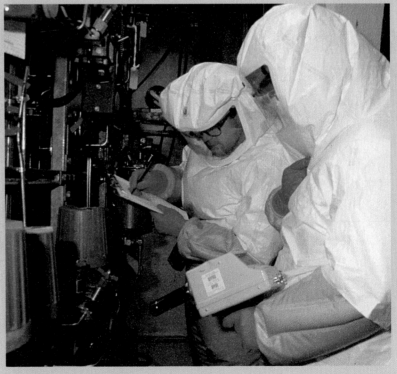

Exercises

Key Words

bedrock storage
calandria
chain reaction
eluate
heavy water
meltdown
moderator
nuclear reactor
probability
risk
uranium oxide
yellow cake

Review

1. What factors led to the development of a nuclear research program in Canada?
2. What two isotopes make up natural uranium?
3. What form of uranium is used as fuel in nuclear reactors?
4. What is the chemical name and formula for the uranium pellets used in nuclear reactors?
5. What are the three components of a nuclear reactor?
6. What determines the type of moderator used in a nuclear reactor?
7. Explain what is meant by a nuclear chain reaction.
8. If the fission reaction ratio is greater than one, what type of moderator should be used and why?
9. Construct a table comparing types of reactors, fuel used, moderator, advantages, and disadvantages.
10. What is the function of a control rod in a nuclear reactor?
11. What does CANDU stand for?
12. Differentiate between heavy water and light water.
13. What makes the nuclear industry's waste problem unique?
14. What is a coagulant?
15. Why is plutonium known as a long-lived thermonuclear battery?

Extension

1. Use your library resource center to research
   a. the incidents that led to and followed the Chernobyl accident
   b. the long-term effects of this accident on the residents of the area

   Report your findings to the class.

Glossary

acid A substance characterized by the presence of free hydrogen ions.

acid rain Rain water that contains gaseous oxides of nonmetals, such as CO_2, SO_2, SO_3, NO_2, dissolved in it.

activity series A table of elements arranged in the order in which they transfer electrons or displace one another from solution.

added ingredients Natural ingredients such as salt, sugar, vitamins, minerals, and spices that are added to foods to improve their nutritional value or add flavor (see **food additive**).

aliphatic compound A hydrocarbon in which the carbon atoms are joined together in a straight chain.

alkali metals The elements found in Group 1 of the periodic table.

alkaline earth metals The elements found in Group 2 of the periodic table.

alkane A saturated hydrocarbon containing only single bonds joining the carbon atoms to each other.

alkene An unsaturated hydrocarbon containing at least one double bond joining two carbon atoms to each other.

alkyne An unsaturated hydrocarbon containing at least one triple bond joining two carbon atoms to each other.

alloy A substance made up of a solution of two or more metals.

alpha decay The type of radioactive decomposition in which a helium nucleus (alpha particle) is emitted from the nucleus.

alpha particle A helium nucleus, composed of two protons and two neutrons.

amino acids The "building blocks" of proteins.

amylase An enzyme that breaks down starch.

analysis A process used to separate and to identify the components of a substance.

anion A negatively charged ion.

anode The electrode in an electrochemical cell at which oxidation occurs.

antacid A base. Substances containing bases are often used to neutralize excess stomach acid.

antibiotic A chemical produced by living organisms which slows or prevents the growth of bacteria.

antiseptic A chemical which slows or stops the growth of germs, but does not kill them on contact.

applied chemistry The branch of applied science that uses chemistry to develop technologies that improve people's lives.

applied science The science that uses scientific knowledge to solve problems.

applied scientist A scientist who takes the laws and theories developed by pure scientists and applies them to some problem or need in society.

aqueous solution A solution made by dissolving a solute in water. Abbreviation: aq.

aromatic compound A hydrocarbon in which the carbon atoms are joined together in a ring with alternating double and single bonds.

artificial transmutation The process of bombarding a nucleus with high-speed particles, which causes the nucleus to emit a proton or neutron to gain stability.

atom The smallest particle of an element that has the properties of that element.

atomic mass The mass of an atom compared to the mass of carbon-12 expressed in atomic mass units, u.

atomic mass number The number of neutrons plus the number of protons in the nucleus of an atom.

atomic number The number of protons in the nucleus of an atom.

atomic radius The estimated distance from the center of the nucleus to the outer edge of the atom.

balanced chemical equation The description of a chemical reaction, using formulas and indicating that the reactants and products have equal numbers of atoms of each type.

barbiturate A drug that is a depressant and acts like a sedative.

base A substance characterized by the presence of free hydroxide ions.

battery One or more electrochemical cells in combination, used to produce energy.

bauxite An ore containing aluminum oxide, Al_2O_3, from which aluminum metal is extracted.

bedrock storage A proposed storage method for radioactive waste. It involves burying the waste in tunnels cut into bedrock, with a clay overlay.

beta decay The type of radioactive decomposition in which an electron (beta particle) is emitted from the nucleus.

beta particle A high-energy electron emitted from a nucleus during radioactive decomposition.

binary acid An acid that contains only two elements.

binary compound A compound that contains only two elements.

biochemistry The chemistry of living things.

biodegradable Capable of being broken down by bacteria into harmless products.

black box Scientific term for an object which is unobservable, or has not been observed.

bleach An unstable chemical substance that decomposes easily and is used to change the color of other substances.

Bohr atom A model of the atom in which electrons occupy distinct energy levels.

Bohr-Rutherford diagram A drawing of an atom that shows electrons in different energy levels.

builders Chemicals added to detergents to soften water and stop scum from forming.

bubble plate Part of a distillation tower used to separate fractions of petroleum.

calandria The part of a CANDU reactor in which fission takes place.

carbohydrate An energy-rich molecule that contains carbon, hydrogen, and oxygen atoms.

carbon dating A process that uses the amount of radioactive carbon-14 in organic matter to measure the age of the matter.

carcinogen A substance that causes cancer.

catalyst A substance that changes the rate of a chemical reaction without being permanently changed itself.

cathode The electrode in an electrochemical cell at which reduction occurs.

cathode ray The invisible ray that originates at the cathode of a cathode ray tube.

cathode ray tube A glass tube containing electrodes in which a beam of cathode rays comes from the cathode. These are the tubes used in television sets and computer screens.

cation A positively charged ion.

chain reaction A reaction in which the neutrons released by the fission of one nucleus trigger fission in other nuclei nearby.

characteristic property A property by which one substance can be distinguished from another.

chemical bond The forces of attraction between atoms in a substance that hold them together.

chemical change A change that produces new substances with new composition and properties.

chemical equation The description of a chemical reaction using formulas.

chemical property A property of a substance that is observed when the substance undergoes a change in composition.

chemical reaction The type of reaction in which the properties of the product(s) are different than those of the reactant(s).

chemical symbol Short form of an element's name.

chemistry The science that deals with the composition, structure, properties, and transformations of matter.

chromatography The separation of components of a mixture according to how rapidly they are transported by a solvent.

cleanser Substance used for cleaning and to kill germs.

coefficient A number placed in front of a molecular formula to indicate how many molecules of the compound are taking part in a reaction.

combustion reaction A chemical reaction involving oxygen in which light and heat are produced.

complete combustion A combustion reaction in which there is enough oxygen that only carbon dioxide and water are produced.

compound A pure chemical substance composed of more than one kind of atom.

concentration A measurement for the quantity of solute dissolved in a certain quantity of solution.

conductivity The property of metals, some metalloids, and ionic solutions that allows an electric current to pass through them.

control A standard used for comparison in an experiment.

corrosion The gradual destruction of a substance by chemical action.

cosmetics Substances applied to a person's face, usually to improve appearance.

covalent bonding The type of chemical bonding that results from the sharing of outer energy-level electrons between atoms.

covalent molecule A molecule in which bonds are formed by shared pairs of electrons.

cracking The process in which the chain of a hydrocarbon is broken into smaller chains.

crystal lattice The three-dimensional shape formed by linking ion pairs.

crystallization A process in which crystals are formed.

cyclic compound A hydrocarbon in which the carbon atoms are joined together in a ring by single bonds.

decomposition reaction A chemical reaction in which one substance breaks down into two or more simpler substances.

depressant A drug that lowers the activity of the central nervous system and relieves tension.

detergent A type of cleanser that differs from soaps in that it is made using an alcohol rather than a fat molecule. Detergents do not form the scum that soaps form.

deuterium An isotope of hydrogen that contains one neutron in its nucleus.

diabetes A condition in which there is a lack of the hormone insulin, which is needed to break down sugar in the body (see **insulin**).

diatomic molecule A molecule of a substance that consists of two atoms.

digestion The process by which the body breaks down food into usable molecules.

dilution The addition of a solvent to a solution to make it less concentrated.

direct combination reaction A chemical reaction in which the reactants are added together to produce the product.

disaccharide A carbohydrate that can be broken apart by hydrolysis to produce two monosaccharides.

disinfectant A chemical substance which kills germs or bacteria on contact.

distillation A separation method that is based on the different boiling points of either a liquid/liquid mixture or a solid/liquid mixture.

double covalent bond The chemical bond that occurs when two atoms share two pairs of electrons between them.

double displacement reaction A chemical reaction that involves the exchange of two elements between reactants.

drug A chemical substance that can affect the functioning of living tissue.

dry cell A commonly used type of electrochemical cell.

ductile Able to be drawn out into wires without breaking.

electricity The flow of electrons in a wire or solution.

electrochemical cell A system that converts the chemical energy of an oxidation-reduction reaction into electrical energy.

electrochemistry Chemical reactions that involve the transfer of electrons.

electrode The part of an electrochemical cell where an oxidation or reduction reaction takes place.

electrolysis The process in which electricity is passed through a substance and causes a chemical change.

electrolyte The substance that dissolves to produce a solution, or the solution itself, that conducts electricity.

electrolytic cell A system that uses electrical energy to produce a chemical reaction.

electron A type of elementary particle that carries one negative charge.

electron dot diagram A diagram in which only the atom's outer energy-level electrons are shown.

electronegativity A measure of the electron-attracting ability of an element.

electroplating The use of electricity to apply a coating of a metal onto another metal.

element A substance that cannot be broken down into simpler substances by chemical means. An element consists of atoms of the same atomic number.

elementary particles The subatomic particles that make up the atom. They are the neutron, proton, and electron.

eluate A solution that has passed through an ion exchange resin.

emulsifying agent A substance that causes immiscible substances to mix, forming an emulsion.

emulsion A liquid mixture where one liquid, usually a fat, is distributed evenly in tiny droplets throughout the other liquids present.

endothermic process A process in which energy is absorbed.

energy levels An imaginary sphere with definite, fixed energy, in which an electron or electrons are believed to exist.

energy value The amount of energy taken in by the body.

enzyme A type of molecule that regulates most of the chemical reactions that take place in living things.

equilibrium The state of balance that exists between substances.

ester A class of organic compounds that is formed by reacting an acid with an alcohol.

evaporation The change of state by which a liquid changes into a gas. Evaporation is used to separate solutes from their solvents.

excited state The state of an atom that has had an extra amount of energy added to the electrons of the atom above their normal energy level.

exothermic process A process in which energy is released.

experiment A carefully planned and controlled set of procedures to test a hypothesis.

family In the periodic table of elements, a vertical group having similar properties (see **group**).

fat An energy-rich compound formed when a glycerol molecule reacts with three fatty acid molecules.

filtrate The liquid that passes through the porous medium in filtration.

filtration A process of separating large particles from a mixture of particles or solution by passing them through a porous medium.

fission The nuclear reaction in which a heavy nucleus splits into new elements and additional neutrons, and produces a large amount of energy.

flame test A test in which a sample is held in a burner flame. The color given off indicates what element is present.

fluorescence A glowing or emitting of light, as in a cathode ray tube.

food additive A substance or its byproduct, including radiation, that may become part of or affect the properties of a food (see **added ingredient**).

forensic science The branch of applied science that examines and analyzes materials to aid in solving crimes.

formula mass The sum of the atomic masses of the atoms that make up a compound.

fractional distillation The process of separating two or more liquids according to the differences in their boiling points.

fuel cell An electrochemical cell that changes chemical energy directly into electrical energy.

fusion The nuclear reaction in which the nuclei of two different hydrogen isotopes fuse.

galvanizing A method used to prevent corrosion by coating one metal with another.

gamma ray A wave-like type of radiation given off by certain kinds of radioactive materials.

gas Matter that does not have a definite shape or volume and will take the shape of a container.

gem A crystal that is valued for its beauty and rarity.

generic name The true name for a compound as opposed to a brand name.

ground state The most stable energy state of an atom.

group A vertical column in the periodic table of elements that have similar properties (see **family**).

Haber process The production of ammonia from hydrogen and nitrogen by using a catalyst and high temperatures and pressures.

half-cell One of two parts of an electrochemical cell. Oxidation occurs in one, reduction in the other.

half-life The period of time in which half of the original radioactive nuclei present in a sample have undergone radioactive decay.

half-reaction That part of a chemical reaction that occurs at one electrode of an electrochemical cell.

Hall-Héroult process A process by which electrolysis is used to produce molten aluminum.

hallucinogen A drug that creates feelings of euphoria and can cause hallucinations.

halogens The nonmetallic elements found in Group 17 of the periodic table.

hard water Water that contains dissolved salts, for example, of magnesium and calcium.

heavy water Water that contains deuterium.

heterogeneous Nonuniform throughout.

homogeneous Uniform throughout.

hydrocarbon A compound that contains only carbon and hydrogen atoms.

hydrogen ion (H^+) A hydrogen atom from which the electron has been removed, leaving a proton. The concentration of hydrogen ions indicates the pH of a solution.

hydrogenation The addition of hydrogen to an unsaturated hydrocarbon.

hydrolysis The decomposition of a substance by the chemical action of water on it.

hydroxide ion (OH^-) A water molecule from which a hydrogen ion (proton) has been removed.

IUPAC International Union of Pure and Applied Chemists, an organization that sets standards for names of elements, chemical symbols, and so on.

immiscible Unable to be mixed. The term is often used with reference to liquids.

incomplete combustion A combustion reaction in which insufficient oxygen is present and products other than water and carbon dioxide are formed.

indicator A substance that reacts with an acid or base to produce a definite color change.

inference A belief based on assumptions.

insulin A hormone produced by the pancreas that breaks down sugar in the body (see **diabetes**).

ions An atom which has gained or lost an electron or electrons to become charged.

ion exchange A process in which one type of ion is exchanged with another. The process takes place on a surface or in a column filled with an ion exchange resin.

ionic bond The attraction between positive and negative ions due to their opposite charges.

ionic compound A compound in which the ions are held together by ionic bonds.

ionic solid A substance in which atoms are held together by ionic bonds.

ionization The process in which an atom is changed into an ion by the removal of an electron.

ionization energy The amount of energy needed to remove an electron from an atom.

ionizing radiation Alpha, beta and gamma radiation; can cause ionization in living tissues.

isomerization The process in which molecules with the same type and number of atoms are bonded together in different arrangements.

isomers Molecules that have the same type and number of atoms but are bonded in different arrangements.

isotopes Atoms of the same element that have different masses.

kilojoule An SI unit of energy. One joule is the amount of energy required to lift a weight of one newton a distance of 1 m.

law In chemistry, a statement made about the regularities found from the observations made about particular systems.

Law of Conservation of Mass The law that states that mass is neither created nor destroyed during a chemical reaction.

Lewis diagram (structure) A diagram in which single lines represent each shared pair of electrons between two atoms.

line spectrum The type of emission spectrum that can be observed by looking at gas tubes with a spectroscope.

lipid A type of nutrient, including fats, oils, and waxes.

liquid Matter that has a definite volume but not a definite shape.

litmus paper Paper that changes color to indicate the presence of an acid or base.

luster The shiny property of metals.

malleable Able to be hammered, rolled out, shaped, or molded.

mass The quantity of matter in an object.

mass spectrometry The process by which the relative masses of atoms and molecules are measured.

matter Any type of substance that has mass and occupies space.

mechanical mixture A heterogeneous mixture with parts that are visibly distinguishable.

meltdown A situation in which a nuclear reactor loses its coolant. As a result, the core melts, releasing radiation.

metal An element that is malleable, ductile, and a good conductor of heat and electricity.

metalloid An element that has both metallic and nonmetallic properties.

metallurgy The science that studies metals, including ways of refining and using them.

minerals Inorganic compounds, some of which are required by the body.

miscible Able to be mixed.

mixture When two or more kinds of matter are mixed and each keeps its own characteristic properties.

model A visual idea that helps one to understand something that cannot be seen.

moderator The substance in a nuclear reactor that slows down the neutrons present.

molar mass The mass of one mole of a substance.

mole The amount of a substance that has the same mass in grams as its atomic or molecular mass in atomic mass units (u).

molecular formula The formula that shows the number of atoms of each element present in a compound.

molecular mass The mass of one molecule of a substance.

molecule The smallest chemical unit of a substance that can exist independently.

monomer A small molecule that is bonded to a large number of other molecules to form a large molecule (see **polymerization**).

monosaccharide The simplest form of a carbohydrate.

multiple covalent bond The covalent bond that forms when elements share more than one pair of electrons.

net cell voltage The sum of the half-reaction voltages in an electrochemical cell.

net equation An equation showing the ions that are involved in a particular chemical reaction.

neutral Having a net result of zero charge as a result of a balance of positive and negative charge.

neutralization The process in which hydrogen ions and hydroxide ions combine to form water molecules.

neutron The elementary particle that carries no electric charge.

noble gases The least reactive elements, found in Group 18 of the periodic table.

nonmetal An element that is dull, brittle, and does not conduct heat or electricity.

nonpolar covalent bond A covalent bond in which the bonding electrons are shared equally between atoms.

nonpolar molecule A molecule in which the bonding electrons are shared equally between atoms or distributed evenly within the molecule.

nuclear atom Rutherford's model of the atom, in which most of the mass of the atom is in a dense core in the center.

nuclear equation The equation that represents changes in the nuclei of atoms.

nuclear power The use of fissionable materials such as uranium to produce heat to generate electricity.

nuclear reactor A system in which radioactive materials are fissioned under controlled conditions to produce energy.

nucleus The positively charged central part of the atom.

nutrient A chemical substance that is used by living things for energy, growth, maintenance, or repair.

observation The act of obtaining information through the senses.

octet The stable grouping of eight electrons in the outer energy level of an atom.

opiate A category of drugs that creates euphoria and tranquillity; includes morphine, codeine, and heroin.

outer energy-level electrons The electrons that participate in chemical reactions.

oxidation A reaction involving the loss of electrons.

oxidation-reduction (redox) reaction A reaction that involves electron transfer and is the result of the combination of an oxidation reaction and a reduction reaction.

oxide The compound formed when an element combines with oxygen.

oxyacid An acid that contains more than two elements.

particle theory A theory, first proposed by Democritus, that all matter is made up of particles called atoms.

period A horizontal row of elements in the periodic table.

periodic property (law) The trend that the properties of the elements repeat regularly as a function of their atomic mass.

periodic table A table in which elements are arranged according to increasing atomic number and recurring properties.

pH A measure of the acidity of a substance.

physical property A property of a substance that can be determined without changing its composition.

plutonium A metallic element which is radioactive; formed as a byproduct of the fissioning of uranium-238.

polar covalent bond A covalent bond that is caused by the unequal sharing of electrons between atoms.

polar molecule A molecule in which electrons are shared unequally between atoms (see **polar covalent bond**).

pollutant A substance that has an adverse effect on living things if present in high enough concentrations.

polyamide A type of polymer used to make various synthetic fibers.

polyester A synthetic fiber of high tensile strength formed by linking many ester molecules together.

polymer A molecule with a very high molecular mass formed by linking many smaller molecules together.

polymerization The process in which smaller molecules of the same substance join to form a large molecule of that substance (see **monomer**).

polysaccharide Starch molecules, formed from long chains of glucose molecules linked together.

postulate A statement proposed as part of a theory.

precipitate An insoluble substance that separates from a solution as a result of a physical or chemical change.

precipitation The process where an insoluble solid separates from a solution.

preservative A substance added to food to prevent bacterial growth.

probability The mathematical likelihood that a certain event will happen.

product A substance, formed as a result of a chemical reaction, whose properties are different than those of any of the reactants.

property A characteristic of a substance.

protein A type of nutrient made up of long chains of amino acids.

proton The elementary particle that carries a one-plus charge.

prototype The first of a kind.

pure science The science that researches the nature of the universe.

pure substance A substance that contains only one kind of matter and thus has fixed properties and composition.

qualitative analysis Chemical detective work; the use of chemical tests to determine non-numerical information, e.g., what elements are present in a sample.

qualitative property A physical or chemical property that cannot be measured and therefore cannot be expressed in numbers.

quantitative property A physical or chemical property that can be measured and thus expressed in numbers.

quantum A specific amount of energy.

radiation Particles and rays which are emitted at high speeds from radioactive materials.

radioactive Emitting radiation.

radioactive decay (decomposition) The emission of a particle from a nucleus that causes the formation of a new element.

radioactivity The release of particles and invisible rays by decomposition of an unstable atomic nucleus.

radioisotope A radioactive isotope of an element.

reactant The starting material involved in a chemical reaction.

reactivity The chemical property of elements which describes their ability to react.

recrystallization The separation technique based on the different solubilities of two or more substances.

redox reaction See **oxidation-reduction reaction**.

reduction A reaction involving the gain of electrons.

relative activity How easily metals lose electrons in comparison to hydrogen, the standard.

residue The solid material that does not pass through filter paper when a solution is filtered.

risk The degree of chance associated with a probability.

roasting In metallurgy, a combustion reaction that removes metals found in ores and recombines them with oxygen to form metallic oxides.

rust Iron oxides, **FeO** and **Fe_2O_3,** formed when iron is oxidized by oxygen in the air.

salt bridge In an electrochemical cell, a glass U-tube containing an ionic solution that lets ions travel between half-cells.

saponification A chemical reaction in which a fat reacts with a base to form a soap molecule and glycerine.

saturated fat A fat in which the carbon atoms are bonded to as many hydrogen atoms as possible.

saturated hydrocarbon A hydrocarbon in which the carbon atoms are bonded to as many hydrogen atoms as possible.

saturated solution A solution that contains as much dissolved solute as it can at any specific temperature.

science An activity that is designed to increase the knowledge of matter and the universe.

scientific method The process of forming and testing hypotheses and interpreting data.

single bond A chemical bond in which one pair of electrons are shared between atoms.

single displacement reaction A reaction in which one element takes the place of another.

skeleton equation An unbalanced chemical equation.

smelting The reaction at high temperatures between metallic oxides and coal, coke or carbon monoxide, to produce metals.

smog (smoke + f**og)** A type of air pollution.

soap A type of cleanser which is made by the chemical reaction between a fat molecule and a base.

solar system model A model of the atom in which electrons orbit the nucleus in the same way that the planets circle the sun.

solid Matter that has a definite shape and volume.

soluble Able to dissolve.

solute The dissolved substance in a solution.

solution A mixture that is homogeneous. It is transparent if the solution is a liquid or gas.

solvent The dissolving medium in a solution.

specific heat capacity The quantity of heat required to raise the temperature of one gram of a substance by one degree Celsius.

spectator ion An ion present in a system that does not take part in a chemical reaction.

spectroscope A device that breaks white light down into its component colors.

spontaneous decomposition A reaction in which a substance breaks down without any outside influence.

stable octet A grouping of eight electrons in the outer energy level of an atom that makes it very stable and unreactive.

standard A substance that is used for comparison; for example, hydrogen is used as the standard of reactivity for metals.

starch A polysaccharide made up of glucose molecules; a primary source of food energy for human beings.

state The physical structure of a compound: solid, liquid, gas, aqueous.

stimulant A drug that affects the central nervous system, elevating mood and increasing heart rate, blood pressure, drive, and energy.

Stock System A system for naming compounds in which the charge on an ion is written in Roman numerals, in parentheses, after the element to which it refers.

structural formula A formula that shows the structure and bonding arrangements of the molecule.

subatomic particles See **elementary particles**.

substitution reaction See **single displacement reaction**.

sugars Simple carbohydrates; a primary source of energy for human beings

surface tension The forces exerted by the molecules in the surface layer of a liquid.

surfactant A substance that changes the conditions at the surface where one substance comes into contact with another.

suspension A mixture in which particles of a liquid or solid are dispersed throughout another liquid but not dissolved.

synthesis reaction A direct combination reaction in which reactants are added together to produce products.

synthetic A substance that is produced in a laboratory rather than from a natural substance.

technological method A series of steps leading from a problem or need to a final product that fulfills that need.

technology Any applied science that develops useful products for society.

temperature inversion A weather phenomenon in which a layer of warm air lies above a layer of cool air, trapping pollutants near ground level.

theory An idea or hypothesis used to explain a set of related observations.

thermal inversion (see **temperature inversion**).

thermoplastics Plastics that cannot be remolded by heating.

thermoset plastics Plastics that can be remolded by heating.

tranquilizer A drug that is a central nervous system depressant, causing the user to be sedated without becoming sleepy.

transition metals The elements found between Group 2 and Group 13 in the periodic table.

triple covalent bond The covalent bond that forms when two atoms share three pairs of electrons.

unsaturated fat A fat that contains one or more double or triple bonds.

unsaturated hydrocarbon A hydrocarbon that contains one or more double or triple bonds.

unsaturated solution A solution that contains less than the maximum amount of dissolved solute that it can hold at a specific temperature.

uranium oxide (UO_2) The fuel used in nuclear reactors (see **yellow cake**).

verdigris A form of copper(II) carbonate that often forms on copper statues and roofs.

vitamins Organic substances required by the body in small amounts.

water hardness See **hard water**.

water pollution The condition where the level of chemicals or microorganisms in a water supply makes the water unusable.

wetting ability A measure of the surface tension in a liquid. Lower surface tension makes the liquid "wetter."

word equation A word description of a chemical reaction listing the reactants and products.

X ray A form of energy produced when high-speed electrons hit a metal target.

yellow cake Common name for uranium oxide pellets, used as fuel for nuclear reactors.

Index

Photo Credits

Key to abbreviations: B (Bottom); T (Top); L (Left); R (Right); C (Center)

pp. 2–3 Eric Meola/The Image Bank; p. 4 (T) The Bettmann Archive, (B) Andy Caulfield/The Image Bank; p. 5 (T) NASA, (B) Sygma/Ron Siddle/Publiphoto; p. 6 NASA; p. 7 Ontario Hydro; p. 9 Richard Megna/Fundamental Photographs, New York; p. 10 Goodyear Canada; p. 11 (T) The Bettmann Archive/UPI, (B) Richard Megna/Fundamental Photographs, New York; p. 12 Derik Murray/The Image Bank; p. 14 A.M. Rosario/The Image Bank; p. 17 Wolf Kutnahorsky; pp. 22–3 Pete Turner/The Image Bank; p. 32 B. Bouflet/Publiphoto; p. 34 Wolf Kutnahorsky; p. 39 Steve Proenl/The Image Bank; p. 40 Phillip Norton/Valan Photos; pp. 42–3 Alfred Pasieka/The Image Bank; p. 44 Peter Grumann/The Image Bank; p. 47 Tom Pantages; p. 48 (L) Burton McNeely/The Image Bank, (R) André Gallant/The Image Bank; p. 51 The Bettmann Archive; p. 54 Michael Melford/The Image Bank; p. 56 The Bettmann Archive/UPI; pp. 58–9 Dr. Mitsuo Ohtsuki/Science Photo Library; p. 62 The Bettmann Archive; p. 65 Patrick Forestier/Sygma/Publiphoto; p. 67 (T) Alvis Upitis/The Image Bank, (B) Tom Pantages; p. 68 Sunnybrook Health Science Centre; p. 69 The Bettmann Archive; p. 72 Al Satterwhite/The Image Bank; pp. 74–5 Michael Dalton/Fundamental Photographs, New York; p. 76 Wolf Kutnahorsky; p. 77 The Bettmann Archive; p. 78 Dr. Roger Bell, University of Maryland (Department of Astronomy); p. 80 Wolf Kutnahorsky; p. 83 Michael Dalton/Fundamental Photographs, New York; p. 88 Tom Pantages; p. 94 Jean-Luc Lebrun/Publiphoto; p. 98 The Bettmann Archive; p. 113 Tom Pantages; p. 116 Richard Megna/Fundamental Photographs, New York; p. 119 Cliff Feulner/The Image Bank; pp. 122–23 Murray Alcosser/The Image Bank; p. 124 Photograph courtesy of Stadium Corporation of Ontario Limited; p. 126 Wolf Kutnahorsky; p. 133 Richard Megna/Fundamental Photographs, New York; p. 135 Wolf Kutnahorsky; p. 141 Tom Pantages; p. 143 Michael Dalton/Fundamental Photographs, New York; p. 145 Paul Silverman/Fundamental Photographs, New York; p. 146 Paul Silverman/Fundamental Photographs, New York; p. 147 (TL) Murray Alcosser/The Image Bank, (TR) Yuri Dojc/The Image Bank, (BL) Tom Pantages, (BR) Tom Pantages; p. 149 (T) Raymond Reuter/ Sygma/Publiphoto, (B) Murray Alcosser/The Image Bank; pp. 152–53 NASA; p. 162 A. Cornu/Publiphoto; p. 167 Wolf Kutnahorsky; p. 173 NASA; p. 174 Teyss/Publiphoto; pp. 176–77 Michael Melford/The Image Bank; p. 178 Richard Megna/Fundamental Photographs, New York; p. 179 Wolf Kutnahorsky; p. 185 Richard Megna/Fundamental Photographs, New York; p. 189 (L) Grafton M. Smith/The Image Bank, (R) Terje Rakke/The Image Bank; p. 191 The Bettmann Archive/UPI/J. Davies; p. 192 William Rivelli/INCO; pp. 194–95 Laurence Hughes/The Image Bank; p. 198 Wolf Kutnahorsky; p. 199 Paul Silverman/Fundamental Photographs, New York;